# Lecture Notes in Economics and Mathematical Systems 655





For further volumes:
http://www.springer.com/series/300

Wim Heijman • Pierre von Mouche
Editors

# New Insights into the Theory of Giffen Goods

Springer

*Editors*

Prof. Dr. Wim Heijman
Wageningen University
Social Sciences
Hollandseweg 1
6706 KN Wageningen
Netherlands
wim.heijman@wur.nl

Dr. Pierre von Mouche
Wageningen University
Social Sciences
Hollandseweg 1
6706 KN Wageningen
Netherlands
pvmouche@yahoo.fr

ISSN 0075-8442
ISBN 978-3-642-21776-0 e-ISBN 978-3-642-21777-7
DOI 10.1007/978-3-642-21777-7
Springer Heidelberg Dordrecht London New York

Library of Congress Control Number: 2011938019



*Cover design*: eStudio Calamar S.L.

Printed on acid-free paper

Springer is part of Springer Science+Business Media (www.springer.com)

*On vous a sans doute souvent demandé à quoi servent les mathématiques et si ces délicates constructions que nous tirons tout entières de notre esprit ne sont pas artificielles et enfantées par notre caprice. Parmi les personnes qui font cette question, je dois faire une distinction: les gens pratiques réclament seulement de nous le moyen de gagner de l'argent. Ceux-là ne méritent pas qu'on leur réponde.*

*Henri Poincaré,*
*La Valeur de la Science, Chapitre V.*

# Preface

This book is about a class of goods which was already identified in the nineteenth century: Giffen goods. One might expect that the elapse of such a period of time would be sufficient to deal with all the aspects of this peculiar type of goods. This, however, appears not to be true. To us, at least for the past three years, the unknown aspects of it have been a source of inspiration. A small group of researchers throughout the world appear to be still active in trying to find answers to the unsolved problems connected to Giffen goods. The book is an attempt to solve a number of them, the majority being of a mathematical nature.

We are grateful to our co-authors for the smooth and fruitful cooperation. Furthermore, the permission granted from Gads Forlag from Copenhagen and Svante Wold, the late Herman Wold's son, to re-publish his father's classical article 'On Giffen's Paradox', first published in the 'Nordisk Tiddskrift for Teknisk Okonomi' in 1948, is gratefully acknowledged here. Lastly, we would like to thank Katharina Wetzel-Vandai, Irene Barrios-Kezic, and Kay Stoll from Springer for their positive part and valuable assistance in the process of publishing this book.

Wageningen,
March 2011

*Wim Heijman*
*Pierre von Mouche*

# Contents

# Contributors

**Erik J. Balder** Mathematical Institute, University of Utrecht, P.O. Box 80010, 3508TA Utrecht, The Netherlands, e.j.balder@uu.nl

**Junko Doi** Faculty of Economics, Kansai University, 3-3-35 Yamate-cho Suita, Osaka, Japan, jkdoi@ipcku.kansai-u.ac.jp

**Rodney J. Garratt** Department of Economics, University of California, Santa Barbara, CA 93106, United States of America, garratt@econ.ucsb.edu

**Rein Haagsma** Wageningen University, Hollandseweg 1, P.O. Box 8130, 6700 EW Wageningen, The Netherlands, rein.haagsma@wur.nl

**Wim Heijman** Wageningen Universiteit, Hollandseweg 1, Postbus 8130, 6700 EW Wageningen, The Netherlands, wim.heijman@wur.nl

**Kazumichi Iwasa** KIER, Kyoto University, Yoshida-Honmachi, Sakyo-Ku, Kyoto 606-8501, Japan, kazumichi@hi-net.zaq.ne.jp

**Kris De Jaegher** Utrecht University School of Economics, P.O. Box 80125, 3508TC Utrecht, Netherlands, k.dejaegher@uu.nl

**Peter G. Moffatt** School of Economics, University of East Anglia, Norwich NR4 7TJ, United Kingdom, p.moffatt@uea.ac.uk

**Pierre H.M. von Mouche** Wageningen University, Hollandseweg 1, P.O. Box 8130, 6700 EW Wageningen, The Netherlands, pvmouche@yahoo.fr

**Willem Pijnappel** Spechtstraat 72, 5932 VK Tegelen, The Netherlands, pijnappel2@yahoo.com

**Jan Rouwendal** Department of Economics, Vrije Universiteit, De Boelelaan 1105, 1081HV Amsterdam, The Netherlands, jrouwendal@feweb.vu.nl

**Koji Shimomura († 2007)** Kobe University, Japan

**Peter N. Sørensen** Department of Economics, University of Copenhagen, Øster Farimagsgade 5 Building 26, 1353 Copenhagen K, Denmark,
peter.sorensen@econ.ku.dk

**Herman O.A. Wold († 1992)** Uppsala University, Sweden

# Introduction

**Wim Heijman and Pierre von Mouche**

This book deals with the neoclassical consumer theory of Giffen goods. According to this theory these are goods for which the quantity demanded increases if their price increases. There are other theories that describe mechanisms for such relationships, leading to, for example, Veblen goods. The following example, being a variant of an example by Marshall that initiated the thinking about Giffen goods, illustrates their plausible existence.

Assume that Robert is planning a long journey by train as comfortably as possible. He has the choice between cheap second-class tickets and expensive first-class tickets. There is a maximum travel budget available which does not cover the cost of first-class train tickets entirely. So, he decides upon a mix of second-class and first-class tickets in such a way that the budget is spent. The day he requests the travel agency to order the tickets it appears that the second-class tickets have become more expensive, so that his chosen mix of second-class and first-class tickets cannot be covered by the budget anymore. What is he to do? The only way of making the journey while keeping the travel costs within the budget is buying more second-class tickets and fewer first-class tickets. Thus second-class tickets is a Giffen good for Robert.

The example considers an individual demand, but neoclassical consumer theory also allows for Giffen goods at the level of market demand. However, empirical evidence at the aggregate level is scarce. Only recently Jensen and Miller have found strong evidence for such Giffen goods: rice for poor consumers in the Hunan province in China. Earlier attempts to demonstrate such existence have been discredited because the data were not correctly interpreted or analysed.

The book contains 11 contributions about Giffen goods. Eight of them, the core of the book, deal with the following question raised by Moffatt: to what extent

W. Heijman (✉) · P. von Mouche
Wageningen Universiteit, Hollandseweg 1, Postbus 8130, 6700 EW Wageningen, The Netherlands
e-mail: wim.heijman@wur.nl; pierre.vanmouche@wur.nl

W. Heijman and P. Mouche (eds.), *New Insights into the Theory of Giffen Goods*, Lecture Notes in Economics and Mathematical Systems 655,
DOI 10.1007/978-3-642-21777-7_1, 

are Giffen goods, for a single consumer, theoretically possible in the neoclassical framework of utility maximisation under a budget restriction? One important characteristic of all contributions is that they aim for mathematical precision.

Already the first contribution of Haagsma, 'Notes on Some Theories of Giffen Behaviour', respects the wish for precise mathematical reasoning. Haagsma deals with the history of the theory on Giffen goods and investigates a number of explanations for Giffen behaviour. In doing so he presents a fresh view of an impressive part of the literature.

The second contribution of Balder, "Exact and Useful Optimization Methods for Microeconomics" shows that there is probably no mathematically rigorous textbook on micro-economics theory. One reason for this lack is that the domain of differentiability of the utility functions is often not correctly identified. Balder presents an efficient solution method by which one can deal with the utility maximisation problem for utility functions with various theoretical properties. Balder's contribution itself does not deal with Giffen goods, but provides the very fundamentals for each theoretical exposition dealing with utility maximisation. Its observations may have far-reaching consequences for the mathematical correctness of various micro-economic articles.

In the literature there are two ways of defining Giffen goods. One for a discrete price change and one for a continuous price change. At first sight one might think that the associated definitions are equivalent. The third contribution "On the Definitions of Giffen and Inferior Goods" by von Mouche and Pijnappel shows that they are not equivalent by providing a non-decreasing demand function that is nowhere strictly increasing. They also deal with the relation between Giffen and inferior goods. It is well-known, that a Giffen good is under weak conditions an inferior good. Von Mouche and Pijnappel make these conditions very weak: by proving a strong form of the compensated law of demand they derive the result that for an upper semi-continuous utility function each Giffen good is inferior.

In the fourth contribution "Giffen Behaviour and Strong Asymmetric Gross Substitutability" De Jaegher investigates, for the case of two goods, other necessary conditions than inferiority for good 1 to be Giffen. He shows that, under reasonable conditions, good 1 is a necessity, good 2 is a luxury, good 1 is a gross substitute for good 2, good 2 is a gross complement to good 1, good 1 is price inelastic and good 2 is price elastic.

Depending on the specific properties the utility function for a Giffen good should have, finding a simple concrete function may not be so easy (if possible at all), although verification that a given function has the desired properties is in general straightforward enough. The situation is similar to finding an antiderivative of a function: finding one may be difficult, but once found its verification is in general easily done. Concrete utility functions are a common theme in contributions three to nine.

The fifth contribution by Heijman and von Mouche "A Child Garden of Concrete Giffen Utility Functions: a Theoretical Review" revisits the concrete utility functions for Giffen goods that have appeared in the literature. Depending on the specific reasonable properties such a function should have, they distinguish between three types of problems: "the text book Giffen problem", the "weak Giffen problem" and the "strong Giffen problem".

The sixth contribution "On Giffen's Paradox" concerns a reprint of an article of Wold from 1948, where he solves the weak Giffen problem, i.e. the problem of providing a concrete continuous, strictly increasing quasi-concave (piece-wise defined) utility function for a Giffen good. This was overlooked by other authors who also tried to solve this problem. In fact Wold's article was almost forgotten in the literature.

In the seventh contribution "Giffen Demand for Several Goods", Sørensen provides, for the setting of $n$ goods, concrete utility functions for which $n - 1$ goods are Giffen; more precisely this is done by fixing the income one and for all. Sørensen does this by providing a class of functions, each of them being a Leontiev utility function with at each entry an appropriate utility function substituted. Each of these utility functions again solves the weak Giffen problem.

As mentioned above, the weak Giffen problem was first solved by Wold. The text book Giffen problem was first solved by Spiegel. Loosely speaking, the latter problem concerns providing a concrete utility function for a Giffen good where the Giffen property can be shown by solving the equation of budget-balancedness together with the equation saying that price ratio equals marginal rate of substitution. The strong Giffen problem concerns providing a concrete utility function for a Giffen good that simultaneously solves the weak and text book Giffen problems.

The eighth contribution "Giffen Behavior Independent of the Wealth Level" concerns a reprint from 2009 by Doi, Iwasa and Shimomura. In this contribution they solve the strong Giffen problem (with a piece-wise defined function). Proving the Giffen property for this function was a real tour de force. Another important contribution in their work is that they show that Giffen behaviour is compatible with an arbitrary low income share for the Giffen good.

As said above, finding a utility function for a Giffen good with specific theoretical properties may be difficult. In the ninth contribution "A Class of Indirect Utility Functions Predicting Giffen Behaviour", Moffatt provides a systematic approach for finding Giffen utility functions with the Giffen property (in the case of two goods). He fixes once and for all income and geometrically investigates what properties an indirect utility function in the two-dimensional price space should have for this and then constructs such functions. Applying Roy's identity to the indirect utility function then provides a Giffen utility function.

The tenth and eleventh contribution do not deal anymore with the traditional strict setting of the utility maximisation under a budget restriction. In the tenth contribution "Close Substitutes and Upward-Sloping Demand Curves" by Rouwendal three alternative circumstances in which Giffen behaviour may arise are presented. These circumstances have in common that "Demand" refers to categories of goods that are close substitutes and "price changes" concern the prices of all goods in a category. The mechanism causing the Giffen behaviour here is different from that of the Giffen goods in the traditional setting.

The compensated law of demand concerns a necessary condition for compensated price changes. In the eleventh contribution "Lotteries and the Law of Demand", Garratt formulates this law for lotteries in situations with a non-convex set of consumption bundles. The finding is that its standard expression necessarily holds in expectation only.

# Notes on Some Theories of Giffen Behaviour

**Rein Haagsma**

**Abstract** Alfred Marshall's introduction of the "Giffen paradox" has kept the minds of many economists occupied for more than a century, as has the more general issue of the possibility of an upward sloping segment of the demand curve. In this paper, I will briefly discuss the main historical roots of this literature and investigate a number of explanations of a positively sloping demand curve.

## 1 Introduction

It was not until the third (1895) edition of his *Principles* that Alfred Marshall [28] stated that the Law of Demand may not always hold. In all likelihood, Marshall was responding to the controversy that had arisen about his consumer surplus analysis during the early 1890s [11, 50], in which he had argued that consumer surplus could provide a practical measure of subjective welfare. His "engine of enquiry" was driven by two assumptions that were heavily debated: (1) that the utility function is additively separable, and (2) that the marginal utility of money is constant. Despite the critics, Marshall insisted on the belief that on many occasions these two assumptions are not unrealistic, particularly in cases of small changes in price. After having explained that it is seldom necessary to take account of changes in the marginal utility of money, he inserted a new section in the third edition with the famous "Giffen paradox":

> There are however some exceptions. For instance, as Mr Giffen has pointed out, a rise in the price of bread makes so large a drain on the resources of the poorer labouring families and raises so much the marginal utility of money to them, that they are forced to curtail their consumption of meat and the more expensive farinaceous foods: and, bread being still the cheapest food which they can get and will take, they consume more, and not less of it. But such cases are rare; when they are met with they must be treated separately (p. 208).

R. Haagsma (✉)
Wageningen University, Hollandseweg 1, P.O. Box 8130, 6700 EW Wageningen, The Netherlands
e-mail: rein.haagsma@wur.nl

W. Heijman and P. Mouche (eds.), *New Insights into the Theory of Giffen Goods*,
Lecture Notes in Economics and Mathematical Systems 655,
DOI 10.1007/978-3-642-21777-7_2, 

It is only in rare cases, according to Marshall, that quantity demanded increases in response to a rise in price – hereafter referred to as "Giffen behaviour". If the price of a commodity that swallows a great part of total expenditure rises, then the induced increase in the marginal utility of money can be so great that consumption of the commodity in question increases.

The precise meaning of this section has kept the minds of many economists occupied for more than a century, as has the more general issue of the possibility of an upward sloping segment of the demand curve. In this paper, I will briefly discuss the main historical roots of this literature and add some comments. Note that Marshall's new section is also inconsistent with his first assumption of additive utility, considering that he accepted the principle of diminishing marginal utility (Gossen's First Law). This is clearly seen if one recalls that the first-order equilibrium condition of constrained utility maximization requires that the marginal utility of bread is equal to its price times the marginal utility of money. Since the right-hand side of this equation increases, the quantity demanded of bread must decrease, rather than increase, under diminishing marginal utility.

Robert Giffen (1837–1910) was a Scottish economic statistician. In 1876, he became head of the statistical department at the Board of Trade and gained a position as chief statistical adviser to the British government. A staunch supporter of free trade, Giffen took a leading part in the public debates on tariff reform during the final decades of the nineteenth century [29]. Notwithstanding his authority and effective public interventions, Giffen would probably have lost his place in the economic canon if Marshall had not mentioned him in the celebrated passage. The question has been raised whether Giffen himself had indeed thought of the possibility of an upward sloping demand curve. Did he find evidence of this? Stigler [43, 44] is quite convinced that in both cases the answer is negative. Which would mean that Marshall gave too much credit to Giffen. Actually it was Simon Gray, employee of the British War Office, who already in 1815 provided a detailed account based on casual observation in a book chapter entitled: "A rise in the price of bread corn, beyond a certain pitch, tends to increase the consumption of it" [17, Book VII, Ch. 5] (see also [30]).

The new section also shed light on another issue. Friedman [15] has pointed out that in the first two editions of the *Principles*, Marshall was curiously vague about whether the demand curve is constructed by keeping money income constant or real income (see also [5]). Was Marshall referring to uncompensated demand or compensated demand? In the third edition, Friedman finds a passage – albeit only one – that puts an end to the confusion. It is the section containing the "Giffen paradox" (together with an added related sentence in Note VI of the Mathematical Appendix). Here clearly an income effect is identified, which offsets a substitution effect. Marshall's example of an upward sloping demand explicitly rules out the interpretation of compensated demand, which ignores the income effect of a price change and therefore is always downward sloping. What caused Marshall to accept the uncompensated version of demand? Friedman notes that the earlier two editions are based on Marshall's groundwork on demand theory which had already been completed in the early 1870s. The third edition was published some 20 years later.

He argues that this period of time was long enough to change Marshall's view under the influence of the publications of Edgeworth and Walras, who advocated the use of uncompensated demand.

Marshall's example considers the basic needs of poor families under subsistence conditions. The pressure of subsistence requirements is also present in the few other texts written by Marshall on the paradox (see Stigler [43]). Some authors have suggested that in these passages Marshall saw consumption decisions not only as being restricted by income constraints, but also by biological constraints regarding minimum nutritional needs. An additional exogenous constraint also appears in Marshall's transportation problem (see Sect. 3), which he mentioned as an alternative "Giffen" example in his correspondence with Edgeworth. Therefore, in this view, Giffen behaviour is just a consequence of outward circumstance. Whether this interpretation correctly sums up Marshall's position or not, after Hicks' work on demand theory, theoretical economists started to look for internal factors. The possibility of a violation of the Law of Demand had to be looked for in a rather peculiar configuration of individual preferences. I will now deal with this modern approach, after which I will return to the case of an additional constraint.

## 2 The Standard Problem

Consider the standard problem of utility maximization under a fixed-income constraint. Hicks and Allen [20] have shown that the price elasticity of demand $\varepsilon$ for some good equals the negative of a weighted average of two preference indices:

$$\varepsilon = -(k\eta + (1-k)\sigma)$$

where weight $k$ is the proportion of income spent on this good (say $X$), $\eta$ is the income elasticity, and $\sigma$ is the elasticity of substitution between this good and a composite good representing all the other goods (say $Y$). It is well-known that income elasticity $\eta$ can be positive, zero, or negative. If it is positive then $X$ is called a normal good and if it is negative then $X$ is called an inferior good. The sign of the income elasticity indicates how the marginal rate of substitution of $X$ for $Y$ changes if the consumer has more of $Y$ for a given quantity of $X$.[1]

[1]This point is not mentioned by Hicks and Allen [20], but may be seen as follows. Suppose two goods $X$ and $Y$ with quantities $x$ and $y$ and a suitable utility function $u(x, y)$. The marginal rate of substitution of $X$ for $Y$ (MRS) equals $u_x/u_y$ (subscripts denote partial derivatives). Differentiating with respect to $y$ gives $(u_y u_{xy} - u_x u_{yy})/u_y^2$. Now, with income $m$, and after some manipulations, differentiation of the first-order conditions of constrained utility maximization gives: $\partial x/\partial m = \lambda(u_y u_{xy} - u_x u_{yy})/\Omega$, where $\lambda$ is the positive marginal utility of money and $\Omega$ is the Hessian determinant of $u$ bordered by $(0, u_x, u_y)$, which also is positive in the case of convex indifference curves. Hence, $\text{sign}(\partial x/\partial m) = \text{sign}(\partial \text{MRS}/\partial y)$. Note that under additive utility, $u_{xy} = 0$, therefore, $\text{sign}(\partial x/\partial m) = \text{sign}(-u_{yy})$ (see below).

The income elasticity is positive (negative) if the marginal rate of substitution strictly increases (decreases) when he has more of $Y$. In a two-goods diagram, a positive (negative) income elasticity is implied by a smaller (larger) vertical distance between indifference curves at higher quantities of $X$ (measured along the horizontal axis). Substitution elasticity $\sigma$ is always a non-negative number and measures the curvature of the indifference curve at the individual's optimum. It varies from zero when the indifference curve forms a right angle – the case of perfect complements – to infinitely large when the indifference curve is flat – the case of perfect substitutes. Note that the price elasticity of the compensated demand equals $-(1-k)\sigma$, which is always negative or zero.

Now it is clear from the above equation that for a positive price elasticity of demand, $X$ must be an inferior good. If a consumer spends a significant part of his income on an inferior good and if the elasticity of substitution between this good and all the other goods is negligible (so $k$ is large and $\sigma$ small), then demand for this good must have an upward sloping segment. This segment is bounded by a descending part at high prices and a descending or ascending part at low prices, as Hicks and Allen show with a diagram (see Fig. 6, p. 68). They argue that Giffen behaviour is perfectly compatible with downward sloping convex indifference curves (which is desired for a constrained utility maximum, rather than minimum), but agree with Marshall that it is only likely under subsistence conditions:

> (...) it is only possible at low levels of income, when a large proportion of expenditure is devoted to this "inferior" commodity, and when, among the small number of other subjects consumed, there are none that are at all easily substitutable for the first. As the standard of living rises, and expenditure becomes increasingly diversified, it is a situation which becomes increasingly improbable (p. 68, 69).

Just like Marshall, therefore, Hicks and Allen appeal to empirical observation to downplay the significance of Giffen behaviour (see also [19, p. 35]). It should be emphasized, however, that there is nothing in the theory itself that makes upward sloping demand "improbable". Economic theory does not tell us that the specific properties of the indifference map, summarised by indices $\eta$ and $\sigma$, only prevail at relatively low levels of utility.[2]

We saw that Marshall believed in the practical value of an additive utility function. By assuming that the marginal utility of money is constant for small changes in price, Marshall showed that, under the principle of diminishing marginal utility, additive utility is consistent with the Law of Demand. Note that his assumption about the marginal utility of money is redundant here: under diminishing marginal utility, an additive utility function necessarily implies that all goods are normal (see footnote 1). The additive utility function, used by Jevons and Walras, became popular in the 1870s. Although Edgeworth proposed the generalized utility function in 1881, the additive form retained its dominant position in economic

[2]See also the comments by Lipsey and Rosenbluth [27], Boland [7] and Dougan [12]. Further note that the weak axiom of revealed preference allows for upward sloping demand [6].

discourse beyond the turn of the century. Pareto even went so far as to state that since the Law of Demand was so well founded in evidence, additive utility has to be a good approximation of preferences [45, p. 327]. In 1915 the Russian economist Slutsky was the first to study the implications of the generalized utility function. Written in Italian, however, the article escaped the attention of English-speaking economists, until it was independently discovered by Schultz and Allen in the 1930s. It was Allen's [1] publication that initially did most to propagate Slutsky's work [2]. Giving a resumé of Slutsky's article, Allen notes that many results in [20] were already explicitly stated by Slutsky. Here we are interested in a result Slutsky derived for the case of an additive utility function.

Slutsky provided the by now familiar argument that the assumption of diminishing marginal utility is not necessary for downward sloping convex indifference curves. In particular, he added that, in the case of additive utility, an appropriate indifference map may also be obtained if one – but only one – good has *increasing* marginal utility[3]:

> (...) If only one of the $u_{ii}$ is positive, the budget is stable if $\Omega > 0$, unstable if $\Omega < 0$. (...) The budget can never be stable if more than one of the $u_{ii}$ is positive [40, p. 46].

Before looking at the implications of this finding, note that in Marshall's time the possibility of increasing marginal utility – where "the appetite comes with eating" – was regarded as somewhat disturbing. Often one tried to bypass the problem by referring to a change in the unit of analysis or a change in preferences [46, p. 395]. Thus Marshall conceded that "the more good music a man hears, the stronger is his taste for it likely to become", but added that, since this all takes time, "the man is not the same at the beginning as at the end of it" (1895, p. 97).

Slutsky found that if one good has increasing marginal utility and all the other goods have decreasing marginal utility, then the former is a normal good and all the other ones are inferior goods (given additive utility and convex indifference curves).[4] For example, with only two goods $X$ and $Y$ and increasing marginal utility for $Y$, it is easily seen that the marginal rate of substitution of $X$ for $Y$ strictly decreases when a consumer has more of $Y$ for a given quantity of $X$.[5] So good $X$ is inferior and good $Y$ is normal. Even, by the Engel aggregation condition, the income elasticity of good $Y$ must be larger than one, so it is a luxury good. Though Slutsky did not further pursue this point, he explicitly recognized the possibility that demand is upward sloping in the case of an inferior good.

For illustration, let us examine a specific additive utility function that gives rise to Giffen behaviour. It is the natural logarithm of the function proposed by Wold

---

[3] See footnote 1 for notation; "budget" refers to a consumer's optimum goods bundle; and "(un)stable" refers to whether the second-order condition is satisfied. See also [1, p. 26].

[4] More recent papers related to this point are Liebhafsky [26] and Silberberg and Walker [39].

[5] See footnote 1.

and Juréen [53, p. 102], who were probably the first to publish a suitable form.[6] Consider the utility function:

$$u(x, y) := \ln(x - 1) - 2\ln(2 - y)$$

with the domain $x > 1$ and $0 \leq y < 2$. Note that the marginal utility of each good is positive. Also, in the case of good $Y$, marginal utility increases as more of it is consumed. Because the indifference curves are strictly convex,[7] this implies that the marginal rate of substitution of $X$ for $Y$

$$-\frac{dy}{dx} = \frac{2 - y}{2(x - 1)}$$

is strictly decreasing in $y$. So $X$ is an inferior good and $Y$ a normal – even luxury – good. With income $m$ and prices $p_x$ and $p_y$, we obtain the demand functions:

$$x = 2 - \frac{1}{p_x}(m - 2p_y)$$

$$y = -2 + \frac{2}{p_y}(m - p_x)$$

provided that income and prices are such that

$$m - 2p_y < p_x < m - p_y.$$

The inequalities ensure that the maximizers lie in the interior of the domain. So suppose these conditions hold and consider the demand function of $X$. As can be seen, demand for $X$ indeed falls as income $m$ rises. The derivative with respect to $p_x$ implies

$$\text{sign}\left(\frac{\partial x}{\partial p_x}\right) = \text{sign}\left(m - 2p_y\right).$$

If $m - 2p_y \leq 0$, then demand for $X$ is defined for all $p_x \in (0, m - p_y)$ and slopes downward (or is constant). If $m - 2p_y > 0$, then demand is defined for all $p_x \in (m - 2p_y, m - p_y)$ and slopes upward.[8] Note that the latter case occurs at

---

[6]The example of Wold and Juréen is based on [52]. It is reproduced as an exercise by Katzner [24, p. 62] and more recently discussed by Weber [49]. There are surprisingly few publications with specific utility functions that imply Giffen behaviour, see e.g. Vandermeulen [47], Silberberg and Walker [39], Spiegel [42], Sörensen [41], Doi et al. [13], and Haagsma [18].

[7]Since the marginal rate of substitution of $X$ for $Y$ is positive, the indifference curves slope downward. The rate of change of their slope is $d^2y/dx^2 = (2 - y)/4(x - 1)^2 > 0$, so the indifference curves also are strictly convex.

[8]The slope of the demand for $Y$ is determined by $\text{sign}(p_x - m)$, which is negative by the right-hand inequality constraint.

relatively *high* incomes, rather than low incomes, in compliance with our earlier remark that there are no a priori reasons to believe that Giffen behaviour can only happen at low standards of living (see also [13, 47]).

## 3 The Case of an Additional Constraint

Let us now consider the case of an additional constraint. The fact that Marshall saw the possibility of an upward sloping demand as a consequence of some additional circumstance is most explicit in a letter written to Edgeworth in 1909. Marshall wrote how a traveller in Holland might choose to increase the distance he goes by canal boat if its fare rose, and decrease his ride on the faster but more expensive train. With this example, Marshall did not so much intend to show the theoretical possibility of an upward sloping demand, but rather, he wanted to convince Edgeworth that this is empirically not improbable [8]. The letter contains the following passage:

> Suppose a man was in a hurry to make a journey of 150 kilos. He had two florins for it, and no more. The fare by boat was one cent a kilo, by third class train two cents. So he decided to go 100 kilos by boat and fifty by train: total cost two florins. On arriving at the boat he found the charge had been raised to $1\frac{1}{4}$ cents per kilo. "Oh: then I will travel $133\frac{1}{3}$ kilos (or as near as may be) by boat, I can't afford more than $16\frac{2}{3}$ kilos by train". Why not? Where is the paradox? [36, p. 441]

It is instructive to formalize Marshall's argument. Suppose a minimum travelling distance of $d$ kilometres. Going by boat involves a speed of $s_x$ km/h and by train $s_y$ km/h. The train is faster: $s_y > s_x$. Budget $m$ and kilometre prices $p_x$ and $p_y$ are such that the minimum distance cannot be travelled by train alone, but only by boat. So it is also assumed that $m/p_y < d < m/p_x$. It follows that the set of feasible options is given by the intersection of the budget set $(p_x x + p_y y \leq m)$ and the minimum distance set $x + y \geq d$, and diagrammatically forms a triangle. Since Marshall's man is "in a hurry", travelling time must be minimized. Therefore, the optimum choice is the feasible option with the most railway kilometres, i.e. the highest $y$. This implies

$$x = \frac{p_y d - m}{p_y - p_x} \text{ and } y = \frac{m - p_x d}{p_y - p_x}$$

(the reader may verify Marshall's numerical assumptions). It is seen that travelling by boat is an inferior activity and that the demand for it indeed slopes upward. Also note that the assumption that travelling by the faster train alone is too expensive is in line with Marshall's observation that Giffen behaviour occurs especially at low levels of income.

How is Marshall's transportation example related to the standard problem of constrained utility maximization? The important point to observe here is that the relative price rule (Gossen's Second Law) is not met. To see this, we may suppose that the man's utility depends negatively on travelling time $t$ according to $u(-t)$.

In Marshall's example, $t = x/s_x + y/s_y$. Then the marginal rate of substitution of boat kilometres for train kilometres is constant and equal to $s_y/s_x$. Because $p_x/p_y < 1 < s_y/s_x$, the marginal rate of substitution is thus greater than the ratio of the kilometre prices. Hence, the transportation example effectively refers to a corner solution where the substitution effect is zero and only a positive income effect remains.

Marshall was not alone in using an additional constraint to illustrate Giffen behaviour. In his *Cours* [35, p. 338], Pareto showed how a poor consumer at subsistence might increase his consumption of the cheapest foods if their price rose, just to stay alive. Wicksell gave a similar example in his *Lectures on Political Economy* [51, p. 60] with a discussion of an individual who exchanges wheat for rye in order to ensure a certain required calorie intake. Both authors gave their examples without any reference to Marshall. Also more recent contributions explain Giffen behaviour by incorporating a biological constraint about minimum nutritional needs (e.g., [10,48]). All these examples have formally much in common. For instance, in both the subsistence case and the transportation case, there is a "superior" good (meat, train kilometres) that contains more of a utility-enhancing characteristic (calories, time saved) than the inferior good (bread, boat kilometres), yet the latter is supplied at a lower price. Also, in both cases, there is a relatively low income budget, and this lack causes a tendency toward specialisation in consumption.

However, the empirical nature of the two constraints differs. The minimum distance constraint is an objective barrier that every person would face under similar circumstances, whereas the subsistence constraint carries a subjective element, also because biological needs are likely to differ from person to person. There is indeed much to say in favour of Davis' [9] view that the subsistence constraint is nothing other than the lowest possible indifference curve (see also [23]). As income falls, consumer preferences will inevitably switch from commodities that are chosen for their taste or social value to commodities that enable one to stay alive, as measured best by their caloric content. When on the brink of starvation, a consumer is willing to substitute one commodity for another only if the caloric loss from the latter is offset by an equivalent caloric gain from the former. This trade-off between goods at rates that just support life is precisely the relationship embodied in the subsistence constraint. Therefore, as Davis concludes, the lowest possible indifference curve must be the subsistence constraint itself.

If the subsistence constraint is seen as an indifference curve, the subsistence explanation of Giffen behaviour changes. When preferences are increasingly motivated by the caloric values of commodities, indifference curves may become linear or, because of nutritional complementarities, even strictly concave. In these cases, the optimum consumption bundle generally implies a set of corner solutions. Now the possibility of Giffen behaviour arises when a price rise moves the optimum choice from an interior position on a convex curve of the indifference map to a corner position on a lower concave curve. Thus in the Marshallian example where initially a poor family consumes both bread and meat, a price rise of bread may force the family to a corner point on a low concave indifference curve where meat consumption is zero and bread consumption is higher than before. This situation is

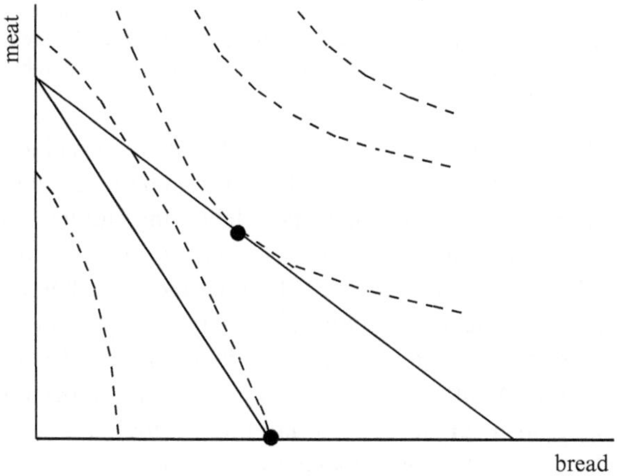


**Fig. 1** Giffen behaviour under a calorie-modified utility function

sketched in Fig. 1 (see also [9]). Note that, as in the additional-constraint case, there is a tendency towards specialisation in consumption; here the tendency even goes all the way.

Summarizing so far then, it can be concluded that the notion of a lowest indifference curve reconciles the two interpretations of the section with the "Giffen paradox": both boil down to a particular configuration of the indifference map. One interpretation allows for concave preferences at the lowest utility levels, based on subsistence conditions that put the caloric content of commodities to the fore. The other interpretation maintains convex preferences, and proposes that Giffen behaviour may obtain in the case of an inferior good with a low elasticity of substitution. Diagrammatically, this implies "elbow-shaped" indifference curves that widen at higher quantities of the inferior good.[9] That such an indifference map can only prevail in some "subsistence zone", as implied by Jensen and Miller (2008), is debatable. For instance, casual observation suggests that many goods, if defined narrowly enough, are inferior for some ranges of income and thus might give rise to Giffen behaviour (see also [5, p. 314]).

## 4 Endogenous Income

Until now we investigated the possibility of upward sloping demand for the case where an individual's income is fixed. Hicks [19] had already discussed the role of the fixed-income constraint when positing the question:

[9]For an example of constructing a specific utility function along these lines, see Moffatt [32] or Sörenson [41].

> What happens if this is not so, if he comes to the market not only as a buyer but also as a seller? Suppose he comes with a fixed stock of some commodity $X$ of which he is prepared to hold back some for his own consumption, if price conditions are favourable to that course of action. (p. 36)

The situation sketched here is characteristic for many smallholder farmers in developing countries, who produce partly for sale and partly for own consumption, and also is not dissimilar from Marshall's poor labouring families. The argument is, however, much more general and extends also to factor markets. In deciding about their labour supply, people act as buyers of leisure time and sellers of working time. Similarly, in their saving decision, people act as buyers of present consumption and sellers of savings deposits. It implies that the price effect will not be the same as when income is fixed. A fall in the price of a good makes a person better off as a buyer but worse off as a seller, thus altering the income effect of a price change. Formally, the Slutsky equation now contains an extra term:

$$\varepsilon = -(k\eta + (1-k)\sigma) + k'\eta$$

where $k'$ is the proportion of income from the endowment of good $X$. Because of the additional endowment income effect, the sign of the overall income effect does not only depend on whether good $X$ is normal or inferior, but also whether $k \lesseqgtr k'$, i.e., whether quantity demanded is smaller or greater than the endowment.

An important implication is that Giffen behaviour can also obtain in case of a normal good. If quantity demanded is much less than the endowment of the good, the income effect is positive and may be so large that this offsets the negative substitution effect of a price change. A well-known example is the backward-bending labour supply curve, or for that matter the forward-bending demand for leisure. When wages are high, labour supply is often significant and so only a few hours of the available daytime – the endowment – are devoted to leisure. A further rise of the wage rate (the shadow price of leisure) may cause an income effect that now outweighs the substitution effect, basically because the satisfaction gained from an extra hour of leisure may be greater than the satisfaction gained from the income earned from working.

Another illustration considers the "Irish potato" – a popular example of actual Giffen behaviour since its early statement in Samuelson's influential textbook *Economics* [38]. Referring to the Great Irish Famine of 1845–1847, and assuming that for many poor families the potato was an inferior good, the example argues that the sharp drop in the supply of potatoes, caused by a fungus on the potato crop, increased the price and indeed also the consumption of potatoes. Although Samuelson's account is not without flaws,[10] largely the same story is told in more

[10] See, e.g., Dwyer and Lindsay [14], Kohli [25], Berg [4], McDonough and Eisenhauer [31], and Rosen [37]. Dwyer and Lindsay make the logical point that, instead of increased, aggregate consumption of potatoes must have decreased, since there were fewer potatoes available for the Irish people. In 1845 about one half of the crop was destroyed and in 1846 about 90%. Less

recent textbooks and publications. Much of this literature discusses the example by treating demand and supply independently. At the time, however, Ireland was very much a self-sufficient peasant economy, where consumers and producers of potatoes were often the same people [14, 37]. Therefore, it is worthwhile to take some time here to study the Irish potato case with the notion of endogenous income.

Suppose an economy with subsistence farmers who make decisions for only two periods, year 1 and year 2. Individual utility depends on the consumption of potatoes in each year, $x_1$ and $x_2$. In the first year, total output of potatoes $m_1$ is used for replanting as seed crop $s$ and consumption:

$$m_1 = s + x_1.$$

Next year's output $m_2$ depends on the size of the seed crop and the net reproduction rate $g$,

$$m_2 = (1 + g)s.$$

Both equations are borrowed from Rosen [37].[11] He mentions that in mid-nineteenth-century Ireland about 15% of the total crop was replanted; a seed potato that yields $1/0.15 \approx 7.5$ potatoes implies a net production rate $g \approx 6.5$. Now, assuming just for simplicity that all second year's output is consumed ($x_2 = m_2$), the intertemporal budget constraint follows as (ignoring discounting)

$$(1 + g)x_1 + x_2 = (1 + g)m_1.$$

It is seen that the relative price of year-1 consumption equals the gross reproduction rate $1+g$. A higher price makes year-1 consumption more expensive, but also raises the return to the amount of seed crops and thus the value of the endowment ($m_1$).

In this simple setting, the potato disease implies a fall in output $m_1$ and a fall in the net reproduction rate $g$. What happens with present consumption of potatoes $x_1$? Of course, a fall in output will reduce or increase this, depending on whether year-1 consumption is normal or inferior. The normal-good case is clearly relevant here, considering the mass starvation and emigration that occurred in Ireland. A fall in the net reproduction rate translates into a lower relative price of year-1 consumption. Noting that $x_1 < m_1$, so quantity demanded is smaller than the endowment, two possibilities arise. If year-1 consumption is inferior, its demand curve slopes downward. A lower price then implies an increase of consumption, which is at odds with the Irish experience. If year-1 consumption is normal, its demand curve may have an upward sloping part. A lower price may then lead to a decrease of consumption. The underlying reason is that the gain in utility from eating an extra potato this year is offset by the gain in utility from using this potato

clear is what happened with potato prices, though Dwyer and Lindsay argue that prices probably increased.

[11] Rosen [37] incorporates these two equations into a larger dynamic model, which he uses to investigate the effects of a transitory and a permanent decline in the productivity of seed potatoes.

for replanting and so having $(1 + g)$ potatoes available for consumption next year. In sum, the Irish potato case is well described by a model in which individuals are both demanders and suppliers of potatoes and the potato is considered to be a normal good.[12]

## 5 Final Remarks

Let me draw to a close with some final remarks. We saw that Marshall's utility theory and his subsistence theory of the Giffen paradox can be reconciled within the same framework: each theory points to a particular configuration of the indifference map. There are, however, a number of "buts", particularly when we want to use this result at the level of markets. First, there is a qualitative difference between individual demand and aggregate demand. Hildenbrand [21] shows that if all individuals have the same demand function and the density of the income distribution is decreasing, then aggregate demand slopes downward, notably even if the commodity is inferior for everyone (see also [3]). Second, the assumption of fixed incomes is questionable. We already found that if individuals have preferences over their factor endowments, an additional income effect appears that may give rise to Giffen behaviour even when the related factor is normal. Moreover, the assumption of fixed incomes is implausible in a general-equilibrium context, where commodity demand generally depends on the returns to factor endowments, such as the wage rate. Nachbar [33] shows that if the equilibrium quantity and the equilibrium price of a commodity both fall in response to an adverse supply shock, then the commodity must have been a normal good.[13]

Recent theories of Giffen behaviour tend to move further away from the standard model of constrained utility maximization. For example, Ng [34] studies what happens if consumers use a so-called step optimization procedure to save decision costs: expenditure is first allocated among a few broad categories (food, housing, etc.) and then distributed within each category. Price changes are assumed to affect expenditures within each category but not the allocation between categories. Consequently, the scope for Giffen behaviour is greatly enlarged for two reasons. First, because the relevant Slutsky equation is now based on only a part of total expenditure, the proportion spent on a good is always higher and the elasticity of substitution between this good and all the other goods of the same category smaller (or equal), as compared with standard optimization. Second, though not mentioned by Ng, the step optimization procedure increases the likelihood of

---

[12]Although the analysis of Rosen [37] is rather different, he also concludes that the Irish potato must have been a normal good. Somewhat similar conclusions based on an approach with endogenous income are derived by Berg [4] and Nachbar [33].

[13]Nachbar assumes that individuals have only preferences over commodities, not over factors endowments.

effective non-income barriers such as the minimum distance constraint in Marshall's transportation problem. Another example is in the article of Garratt [16], which employs a remark by Hicks about expensive indivisible goods. Garratt shows that the purchase of such a good can cause Giffen behaviour with respect to some divisible good. A fall in the price of a divisible good may induce a consumer to buy an expensive indivisible good, which may lead to a lower quantity demanded of the divisible good. Finally, Hoy and Robson [22] introduce the role of imperfect information. They demonstrate that an individual may reduce his demand for insurance in response to a lower premium when his coefficient of relative risk aversion is greater than one.

All in all, the preceding analysis suggests that Giffen behaviour is not as rare as often claimed, once it is accepted that the standard fixed-income interpretation generally is too strict and also implausible. One implication is that the empirical search for the Holy Giffen example has been misguided too long by this interpretation.

**Acknowledgements** The author thanks the editors of this book and Hans-Peter Weikard for their helpful comments on an earlier draft of the article.

## References

1. Allen, R.G.D., 1936, Professor Slutsky's theory of consumers' choice, *Review of Economic Studies* 3(2), 120–129.
2. Barnett, V., 2004, E.E. Slutsky: mathematical statistician, economist, and political economist?, *Journal of the History of Economic Thought* 26(1), 5–18.
3. Baruch, S. and Y. Kannai, 2001, Inferior goods, Giffen goods, and Shochu, in: G. Debreu, W. Neuefeind and W. Trockel, *Economic essays – A Festschrift for Werner Hildenbrand*, Springer, Berlin.
4. Berg, M., 1987, Giffen's paradox revisited, *Bulletin of Economic Research* 39(1), 79–89.
5. Blaug, 1997, *Economic theory in retrospect*, 5th edition, Cambridge University Press, Cambridge.
6. Boland, L.A., 1975, The Law of Demand, weak axiom of revealed preference and price-consumption curves, *Australian Economic Papers* 14, 104–119.
7. Boland, L.A., 1977, Giffen goods, market prices and testability, *Australian Economic Papers* 16, 72–85.
8. Creedy, J., 1990, Marshall and Edgeworth, *Scottish Journal of Political Economy* 37(1), 18–39.
9. Davis, J.E., 1994, Giffen goods, the survival imperative, and the Irish potato culture, *Journal of Political Economy* 102(3), 547–565.
10. Dooley, P.C., 1988, Porter's hint and alternative theories of the Giffen paradox, *Australian Economic Papers* 27, June,142–144.
11. Dooley, P.C., 1992, Why did Marshall introduce the Giffen paradox?, *History of Political Economy* 24(3), 749–752.
12. Dougan, W.R., 1982, Giffen goods and the Law of Demand, *Journal of Political Economy* 90(4), 809–815.
13. Doi, J., K. Isawa, and K. Shimomura, 2009, Giffen behavior independent of wealth level, *Economic Theory* 41, 247–267.
14. Dwyer, G.P. and C.M. Lindsay, 1984, Robert Giffen and the Irish potato, *American Economic Review* 74(1), 188–192.

15. Friedman, M., 1949, The Marshallian demand curve, *Journal of Political Economy* 57(6), 463–495.
16. Garratt, R., 1997, Indivisibilities, inferior goods, and Giffen goods, *Canadian Journal of Economics* 30(1), 246–251.
17. Gray, S., 1815, *The happiness of states: Or an inquiry concerning population, the modes of subsisting and employing it, and the effects of all on human happiness*, London, Hatchard.
18. Haagsma, R., 2010, A convenient utility function with Giffen behaviour, Unpublished Manuscript, Wageningen University, The Netherlands.
19. Hicks, J.R., 1939, *Value and capital. An inquiry into some fundamental principles of economic theory*, 2nd edition, Oxford, Clarendon press.
20. Hicks, J.R. and R.G.D. Allen, 1934, A reconsideration of the Theory of Value: Part I, *Economica* 1(1), 52–76.
21. Hildenbrand, W., 1983, On the "Law of Demand", *Econometrica* 51(4), 997–1019.
22. Hoy, M. and A.J. Robson, 1981, Insurance as a Giffen good, *Economics Letters* 8, 47–51.
23. Jensen, R.T. and N.H. Miller, 2008, Giffen behavior and subsistence consumption, *American Economic Review* 98(4), 1553–1577.
24. Katzner, D.W., 1970, *Static demand theory*, Macmillan Company, New York.
25. Kohli, U., 1986, Robert Giffen and the Irish potato: Note, *American Economic Review* 76(3), 539–542.
26. Liebhafsky, H.H., 1969, New thoughts about inferior goods, *American Economic Review* 59(5), 931–934.
27. Lipsey, R.G. and G. Rosenbluth, 1971, A contribution to the New Theory of Demand: a rehabilitation of the Giffen good, *Canadian Journal of Economics* 4(2), 131–163.
28. Marshall, A., 1895, Principles of Economics. An introductory volume, 3rd edition, Macmillan, London.
29. Mason, R., 1996, Robert Giffen and the Tariff Reform Campaign, 1865–1910, *Journal of European Economic History* 25(1), 171–188.
30. Masuda, E. and P. Newman, 1981, Gray and Giffen goods, *Economic Journal* 91, 1011–1014.
31. McDonough, T. and J. Eisenhauer, 1995, Sir Robert Giffen and the Great Potato Famine: A discussion of the role of a legend in neoclassical economics, *Journal of Economic Issues* 29(3), 747–759.
32. Moffatt, P.G., 2002, Is Giffen behaviour compatible with the axioms of consumer theory?, *Journal of Mathematical Economics* 37, 259–267.
33. Nachbar, J.H., 1998, The last word on Giffen goods?, *Economic Theory* 11, 403–412.
34. Ng, Y-K, 1972, Step-optimization, secondary constraints, and Giffen goods, *Canadian Journal of Economics* 5(4), 553–560.
35. Pareto, V., 1896, *Cours d 'économie politique*, Rouge, Lausanne.
36. Pigou, A.C., ed., 1925, *Memorials of Alfred Marshall* 1842–1924, Macmillan, London.
37. Rosen, S., 1999, Potato paradoxes, *Journal of Political Economy* 107(6), S294-S313.
38. Samuelson, P.A., 1948, Economics: An introductory analysis, McGraw-Hill, New York.
39. Silberberg, E. and D.A. Walker, 1984, A modern analysis of Giffen's paradox, *International Economic Review* 25(3), 687–694.
40. Slutsky, E.E., 1956, On the theory of the budget of the consumer, in: G.J. Stigler and K.E. Boulding, *Readings in price theory*, George Allen & Unwin, London.
41. Sörensen, P.N., 2007, Simple utility functions with Giffen demand, *Economic Theory* 31, 367–370.
42. Spiegel, U., 1994, The case of a "Giffen good", *Journal of Economic Education 25,* Spring, 137–147.
43. Stigler, G.J., 1947, Notes on the history of the Giffen paradox, *Journal of Political Economy* 55(2), 152–156.
44. Stigler, G.J., 1948, A reply, *Journal of Political Economy* 56(1), 61–62.
45. Stigler, G.J., 1950a, The development of utility theory: I, *Journal of Political Economy* 58(4), 307–327.

46. Stigler, G.J., 1950b, The development of utility theory: II, *Journal of Political Economy* 58(5), 373–396.
47. Vandermeulen, D.C., 1972, Upward sloping demand curves without the Giffen paradox, *American Economic Review* 62(3), 453–458.
48. Van Marrewijk, C. and P.A.G. van Bergeijk, 1990, Giffen goods and the subsistence level, *History of Political Economy* 22(1), 145–148.
49. Weber, C.E., 1997, The case of a Giffen good: Comment, *Journal of Economic Education* 28, Winter, 137–147.
50. White, M.V., 1990, Invention in the face of necessity: Marshallian rhetoric and the Giffen good(s), *Economic Record*, March, 1–11.
51. Wicksell, K., 1934, *Lectures on Political Economy*, Trans. E. Classen and edited by L. Robbins, London, Routledge.
52. Wold, H., 1948, On Giffen's paradox, *Nordisk Tiddskrift for Teknisk Ökonomi* 12, 283–290.
53. Wold, H. and L. Juréen, 1953, *Demand analysis. A study in econometrics*, John Wiley & Sons, New York.

# Exact and Useful Optimization Methods for Microeconomics

**Erik J. Balder**

**Abstract** This paper points out that the treatment of utility maximization in current textbooks on microeconomic theory is deficient in at least three respects: breadth of coverage, completeness-cum-coherence of solution methods and mathematical correctness. Improvements are suggested in the form of a Kuhn-Tucker type theorem that has been customized for microeconomics. To ensure uniqueness of the optimal solution *stringent quasiconcavity*, an apparently new adaptation of the notion of strict quasiconcavity, is introduced. It improves upon an earlier notion formulated by Aliprantis, Brown and Burkinshaw. The role of the domain of differentiability of the utility function is emphasized. This is not only to repair a widespread error in the microeconomic literature but also to point out that this domain can be chosen sensibly in order to include the maximization of certain nondifferentiable utility functions, such as Leontiev utility functions. To underscore the usefulness of the optimality conditions obtained here, five quite different instances of utility maximization are completely solved by a single coherent method.

## 1 Introduction

In the currently popular textbooks [8, 10, 11, 13] on microeconomic theory and mathematical economics, all of which profess to be rigorous and precise, the treatment of the fundamental subject of utility maximization would seem to show considerable shortcomings in the following three respects: breadth of coverage, completeness-cum-coherence of solution methods and mathematical correctness. Similar shortcomings also show up in related textbooks with a less formal orientation [9, 14]. These shortcomings are briefly reviewed and explained in Remark 2.

---

E.J. Balder (✉)
Mathematical Institute, University of Utrecht, P.O. Box 80010, 3508TA Utrecht, The Netherlands
e-mail: e.j.balder@uu.nl

W. Heijman and P. Mouche (eds.), *New Insights into the Theory of Giffen Goods*, Lecture Notes in Economics and Mathematical Systems 655,
DOI 10.1007/978-3-642-21777-7_3, 

As a consequence, the reader will search in vain in the standard literature on microeconomic theory and mathematical economics for a method to derive the Marshallian demand function that meets the following criteria: it must be coherent, complete and correct (i.e., based on generally accepted principles of correct mathematical reasoning starting from, say, the Kuhn-Tucker and Weierstrass' theorems), and it must be operationally useful by being applicable to at least the following five basic instances of standard utility functions, where $\ell \in \mathbb{N}$ stands for the number of commodities: $(i)$ Cobb–Douglas utility function, $\ell > 2$, $(ii)$ CES utility function (more precisely, this instance will be separated below into three different cases $(iia)$, $(iib)$ and $(iic)$), $\ell > 2$, $(iii)$ linear utility function with positive coefficients, $\ell \geq 2$, $(iv)$ the utility function $u(x_1, x_2) = x_1^2(x_2 + 1)$ or any similar one leading to partial corner point solutions, and $(v)$ Leontiev utility function, $\ell \geq 2$. The actual state of affairs in the standard literature is even more perplexing than stated above: the problems signaled in Remark 2 already affect the classical instances $(i)$–$(ii)$ and already for $\ell = 2$. The appendix offers a more detailed account of this.

To motivate the above five instances, I first observe that there are very good reasons for including instances with more than two commodities in the preceding list. Indeed, although not many of the aforementioned references point this out, the standard utility maximization problem in microeconomics (see (1)) can immediately be reduced to an optimization problem over an interval if $\ell = 2$. This follows from the fact that its optimal solution must be budget-balanced; cf. Theorem 1$(a)$. Now an interval optimization problem is quite elementary. One can solve it very directly by means of sign diagrams of the derivative, possibly supported by the use of computer algebra packages. This explains why I take $\ell > 2$ in instances $(i)$–$(ii)$: it is a minimal check of operational usefulness. Next, instances $(iii)$–$(iv)$ require the solution method to be able to detect multiple solutions and corner point solutions. The latter form a standard subject in intermediate microeconomics courses [4]. Remarkably, even though at least three of the above-mentioned textbooks indicate the possibility of corner point or multiple solutions in the form of illustrations (see Fig. 3.D.4 in [10], Fig. 1.10 in [8] and Fig. 1.17 of [13]), none of them manages to treat this subject in an analytically satisfactory and complete manner. Finally, instance $(v)$ has a nondifferentiable utility function, which presents a quite unusual operational challenge for the solution method. I observe that in [8, 10, 11, 13] this Leontiev instance is not included in the mainstream approach.

In Sect. 3 I shall develop an efficient solution method that meets all of the above criteria; it is based on optimality results for general utility maximization that are presented in Sect. 2. This development started in my Utrecht lecture notes, because I was dissatisfied with what the standard microeconomics literature had to offer. Essentially, the optimality results take the form of a Kuhn-Tucker theorem, Theorem 1, that has been customized for microeconomics. To a high degree such customization depends on a special property of utility functions that is commonly found in microeconomics: they are strictly increasing. I observe beforehand that the classical Cobb–Douglas and CES instances $(i)$–$(ii)$ require that the model can deal with utility functions that are defined on the nonnegative orthant $\mathbb{R}^\ell_+$ only, possibly with nondifferentiability on the boundary $\mathbb{R}^\ell_+ \setminus \mathbb{R}^\ell_{++}$. Together with the applicability of my method to the nondifferentiable instance ($v$), this indicates that careful consideration of the differentiability domain of the utility function plays an important role in this paper (cf. Remark 2). Parts ($a$) to ($c$) of the customized Kuhn-Tucker theorem come close to what is done on [8, p. 23 ff.], but, as my applications to instances ($i$), ($iia$)–($iib$) and ($iv$) will show in particular, it is part ($d$) of Theorem 1, absent in [8], which makes a considerable difference. This part uses an apparently new adaptation of the notion of strict quasiconcavity, called *stringent quasiconcavity*, which improves upon a related earlier notion by Aliprantis, Brown and Burkinshaw [2] (see Remark 3). It allows the method, in its handling of sufficient conditions for optimality and uniqueness, to go beyond interior optimal solutions and makes it possible to derive corner point solutions in a rigorous, coherent and efficient way, for instance in instance ($iv$), whose utility function is *not* strictly quasiconcave on $\mathbb{R}^2_+$.

In sum, this paper improves on the usual literature on utility maximization in microeconomics by presenting a custom-made Kuhn-Tucker theorem that exploits the usual strict monotonicity of utility functions in microeconomics via the new notion of stringent quasiconcavity and that pays careful attention to the utility function's domain of differentiability.

## 2 Customized Optimality Results for Microeconomics

Let $u : \mathbb{R}^\ell_+ \to \mathbb{R}$ be a continuous function, the *utility function*. Let $\Omega$ be an open set that is contained in $\mathbb{R}^\ell_+$ (whence in the strictly positive orthant $\mathbb{R}^\ell_{++}$); the function $u$ is supposed to be differentiable on $\Omega$ [6]. In Sect. 3 I shall choose for $\Omega$ the strictly positive orthant $\mathbb{R}^\ell_{++}$ so as to treat instances ($i$)–($iv$), but to deal with the Leontiev instance ($v$) I am going to choose $\Omega$ differently. Throughout I suppose that $u$ is *strictly increasing* on $\mathbb{R}^\ell_+$; that is to say, for every $x$ and $x'$ in $\mathbb{R}^\ell_+$ the following must hold: if $x_i > x'_i$ for every $i = 1, 2, \ldots, n$, then $u(x) > u(x')$. For $p \in \mathbb{R}^\ell_{++}$ (*price vector*) and $y \in \mathbb{R}_+$ (*income*) the consumer's *utility maximization problem* is as follows:

$$\text{maximize } u(x) \text{ over all } x \in \mathbb{R}^\ell_+ \text{ such that } p \cdot x \leq y \tag{1}$$

and it is desired to determine *all* global optima (if any). The problem is well-defined because $0 \in B$. Here $B := \{x \in \mathbb{R}^\ell_+ : p \cdot x \leq y\}$ stands for the feasible set of this problem, which is called the *budget set*. Of course, for $y = 0$ the problem is trivial, because then $B = \{0\}$. I shall also use $B_0 := \{x \in \mathbb{R}^\ell_+ : p \cdot x = y\}$ to denote the so-called *budget plane*. A vector $x$ in $\mathbb{R}^\ell_+$ is said to be *budget-balanced* if $p \cdot x = y$, that is to say, if it belongs to the budget plane $B_0$. Recall that $u : \mathbb{R}^\ell_+ \to \mathbb{R}$ is defined to be *quasiconcave* on $\mathbb{R}^\ell_+$ if $u(tx + (1-t)x') \geq \min(u(x), u(x'))$ for every pair $x, x' \in \mathbb{R}^\ell_+$ and every $t \in [0, 1]$. I shall say that the function $u : \mathbb{R}^\ell_+ \to \mathbb{R}$ is *stringently quasiconcave* if it is both quasiconcave on $\mathbb{R}^\ell_+$ and if it has the following property, which I shall call *property* ($S$): for every pair $x, x' \in \mathbb{R}^\ell_+$ with $x \neq x'$ and $u(x) = u(x') > u(0)$ one has $u(\frac{1}{2}x + \frac{1}{2}x') > u(x) = u(x')$.

It is worth noting that in the main text of this paper the stringent quasiconcavity notion is *always* considered on the full domain $\mathbb{R}^{\ell}_{+}$ of the utility function $u$ (but see Remark 1$(ii)$). Observe that stringent quasiconcavity is a modification of the classical notion of strict quasiconcavity which excludes certain points that are sub-optimal (if $y > 0$) and only works with points $x$ and $x'$ at the same utility level. Clearly, a sufficient condition for $u : \mathbb{R}^{\ell}_{+} \to \mathbb{R}$ to be stringently quasiconcave is that it is quasiconcave on $\mathbb{R}^{\ell}_{+}$ and strictly quasiconcave when restricted to the set $C := \{x \in \mathbb{R}^{\ell}_{+} : u(x) > u(0)\}$ (i.e., if $u(tx + (1-t)x') > \min(u(x), u(x'))$ for every pair $x, x' \in C$, $x \neq x'$, and every $t \in (0,1)$). A related sufficient condition for stringent quasiconcavity is given below in Proposition 1; it shows the utility function to be stringently quasiconcave in some of the previously mentioned instances. Because it turns out to work so very well for utility maximization, it is surprising that the stringent quasiconcavity notion was neither introduced before in economics nor implicitly exploited in another guise. Yet this appears to be the case.

**Theorem 1.** $(a)$ *The consumer's utility maximization problem (1) has an optimal solution. Moreover, every optimal solution is budget-balanced.*
$(b)$ *Suppose that (1) has an optimal solution $x^*$ which is such that $x^* \in \Omega$. Then there exists $\lambda \geq 0$ such that*

$$\nabla u(x^*) = \lambda p. \tag{2}$$

$(c)$ *If $x^* \in \Omega$ is budget-balanced and such that (2) holds for some $\lambda > 0$, then $x^*$ is an optimal solution of (1), provided that u is quasiconcave on $\mathbb{R}^{\ell}_{+}$.*
$(d)$ *If u has property $(S)$, then (1) has a unique optimal solution. In particular, if u is stringently quasiconcave, then any budget-balanced $x^* \in \Omega$ for which (2) holds for some $\lambda > 0$, is the unique optimal solution of (1).*

This result has the familiar makeup of results in optimization theory: existence, followed by necessary conditions for optimality that are sharpened into sufficient conditions and even a uniqueness condition. Simple examples show that the above formulation is sharp; for instance, taking $\ell = 1$, $u(x) := (x-1)^3$, $y = 1$ and $p = 1$ shows that the possibility $\lambda = 0$ cannot be excluded in part $(b)$, etc.

**Lemma 1.** *Suppose that u is quasiconcave on $\mathbb{R}^{\ell}_{+}$. Then for every $x \in \Omega$ and $x' \in \mathbb{R}^{\ell}_{+}$*

$$u(x) \leq u(x') \text{ implies } \nabla u(x) \cdot (x' - x) \geq 0.$$

*Proof.* For $t \in [0,1]$ let $\phi(t) := u((1-t)x + tx')$. Then $u(x) \leq u(x')$ implies $\phi(t) \geq u(x) = \phi(0)$. So $\phi(t)$ attains a minimum over $[0,1]$ for $t = 0$. Also $\phi$, the composition of $u$ and a linear mapping, is differentiable from the right at 0, because $u$ is differentiable at $x \in \Omega$. It follows that $\phi'(0) \geq 0$, as a first order necessary condition for optimality. The desired inequality then follows by the chain rule. □

*Proof.* Now Theorem 1 can be proved.

($a$) Existence of an optimal solution $x^*$ follows by the Weierstrass theorem, because $u$ is continuous and $B$ is a nonempty compact set. For $y = 0$ the identity $B = B_0 = \{0\}$ causes budget balancedness to hold trivially. For $y > 0$, $x^* \notin B_0$ would imply $p \cdot x^* < y$. So setting $\tilde{x}_i := x_i^* + t$ for $i = 1, \ldots, \ell$ would result in a contradiction for $t > 0$ sufficiently small, because then $\tilde{x} \in B$ and $u(\tilde{x}) > u(x^*)$ by the fact that $u$ increases strictly.

($b$) The hypothesis $x^* \in \Omega$ implies that $x^*$ is also an optimal solution of the auxiliary optimization problem

$$\text{maximize } u(x) \text{ over all } x \in \Omega \text{ with } p \cdot x \leq y,$$

which has only one inequality constraint. At this stage it is traditional to invoke a version of the Kuhn-Tucker theorem and preferably – to avoid incomplete or inaccurate arguments – a version that allows open domains of definition for its functions: e.g., Theorems 1.D.3–1.D.4 in [13] can be used. However, the following extremely simple argument can be used instead, based on budget-balancedness in Theorem 1($a$) (incidentally, a different but similarly simple argument, based on polar considerations, can also be given for the consumer's expenditure minimization problem to prove a result that is very similar to Theorem 1). By part ($a$) the bundle $x^*$ is also an optimal solution of the auxiliary optimization problem

$$\text{maximize } u(x) \text{ over all } x \in \Omega \text{ with } p \cdot x = y.$$

Here the constraint can be expressed as $x_\ell = (y - \sum_{i=1}^m p_i x_i)/p_\ell$, where $m := \ell - 1$. Hence, the previous optimization problem can be rephrased in the variables $x_1, \ldots, x_m$ as follows

$$\text{maximize } \tilde{u}(x_1, \ldots, x_m) := u\left(x_1, \ldots, x_m, \frac{y - \sum_{i=1}^m p_i x_i}{p_\ell}\right) \text{ over all}$$

$$(x_1, \ldots, x_m) \in \tilde{\Omega}.$$

Here $\tilde{\Omega}$ is the set of all $(x_1, \ldots, x_m) \in \mathbb{R}^m$ such that $(x_1, \ldots, x_m, (y - \sum_{i=1}^m p_i x_i)/p_\ell) \in \Omega$. By continuity of linear maps it is clear that $\tilde{\Omega}$ is open. Because $\tilde{u}$ attains its maximum over the open set $\tilde{\Omega}$ at $\tilde{x}^* := (x_1^*, \ldots, x_m^*)$, it follows from standard calculus results and the chain rule that $0 = \partial \tilde{u}(\tilde{x}^*)/\partial x_i = \partial u(x^*)/\partial x_i - p_i p_\ell^{-1} \partial u(x^*)/\partial x_\ell$ for $i = 1, \ldots, \ell - 1$. So setting $\lambda := p_\ell^{-1} \partial u(x^*)/\partial x_\ell$ gives (2).

($c$) Suppose that $x^* \in B_0 \cap \Omega$ were not optimal. Then there would be $\hat{x} \in B$ such that $u(\hat{x}) > u(x^*)$. So $u((1-t)\hat{x}) > u(x^*)$ would hold for $t > 0$ small enough, by continuity of $u$. By Lemma 1 the quasiconcavity hypothesis for $u$ implies

$$\nabla u(x^*) \cdot (x - x^*) \geq 0 \text{ for every } x \in \mathbb{R}_+^\ell \text{ with } u(x) \geq u(x^*).$$

By $\nabla u(x^*) = \lambda p$ and $\lambda > 0$, it then follows that $p \cdot ((1-t)\hat{x} - x^*) \geq 0$, whence $p \cdot \hat{x} > (1-t)p \cdot \hat{x} \geq y$ by the given budget-balancedness of $x^*$ (observe that $p \cdot \hat{x} > 0$ by $\hat{x} \neq 0$). This contradicts $\hat{x} \in B$.

($d$) First, if $y = 0$ then $B = \{0\}$, so $x^* = 0$ is the unique optimal solution. Next, if $y > 0$, then $u(x^*) > u(0)$ must hold for any optimal solution $x^*$, because $(t, t, \ldots, t)$ belongs to $B$ for $t > 0$ small enough and because $u$ is strictly monotone. Now suppose that $x^*$ and $x^{**}$ were two different optimal solutions of (1). Then $u(x^*) = u(x^{**}) =$ optimal value of (1) and $u(x^*) > u(0)$ by the previous argument. Define $\hat{x} := \frac{1}{2}x^* + \frac{1}{2}x^{**}$; then $\hat{x} \in B$ and property ($S$) gives $u(\hat{x}) > \frac{1}{2}u(x^*) + \frac{1}{2}u(x^{**}) = u(x^*)$. This contradicts the optimality of $x^*$. So the optimal solution is unique. The final part of the statement is an immediate consequence of combining part ($c$) with uniqueness. □

*Remark 1.* ($i$) Without monotonicity in Theorem 1, the budget-balancedness of the optimal solution can obviously not be maintained, but a fair part of Theorem 1 continues to hold when $u$ is nondecreasing and the details are as follows. In part ($a$) the existence of at least one budget-balanced optimal solution is still guaranteed and part ($b$) continues to hold as stated. Part ($c$) of Theorem 1 remains meaningful by linking the additional possibility $x^* \in B \backslash B_0$ with the multiplier $\lambda = 0$. In that case the optimality of $x^*$ can be guaranteed if $u$, next to being nondecreasing, is concave on $\mathbb{R}^\ell_+$ (the example $\ell = 1$, $x^* = 1$, $u(x) := (x-1)^3$, $y = 2$ and $p = 1$ shows that mere quasiconcavity is insufficient in this situation). Finally, part ($d$) is without significance: to have property ($S$), a nondecreasing $u$ must be strictly increasing, a situation that is already covered by Theorem 1($d$) itself.

($ii$) Some utility functions have $\mathbb{R}^\ell_{++}$ as their natural domain of definition. Let $u : \mathbb{R}^\ell_{++} \to \mathbb{R}$ be such a function and suppose that $u$ is continuous and strictly increasing on $\mathbb{R}^\ell_{++}$, as well as differentiable on some open set $\Omega \subset \mathbb{R}^\ell_{++}$. To avoid trivialities, the new situation requires $y > 0$. Then parts ($b$)–($d$) of Theorems 1 continue to hold, naturally with quasiconcavity on $\mathbb{R}^\ell_+$ replaced by quasiconcavity on $\mathbb{R}^\ell_{++}$ and with property ($S$) redefined as follows: for every pair $x, x' \in \mathbb{R}^\ell_{++}$ with $x \neq x'$ and $u(x) = u(x')$ one has $u(\frac{1}{2}x + \frac{1}{2}x') > u(x) = u(x')$. An obvious sufficient condition for these two properties to hold is that $u$ be strictly quasiconcave. As shown by the example $\ell = 2$, $u(x_1, x_2) := x_1 + x_2$, part ($a$) of Theorem 1 needs adjustment: an optimal solution is guaranteed to exist under the extra condition that the set $C_v := \{x \in \mathbb{R}^\ell_{++} : u(x) \geq v\}$ is closed for every $v \in u(\mathbb{R}^\ell_{++})$. Namely, given $p$ and $y > 0$, fix any $\bar{x} \in \mathbb{R}^\ell_{++}$ with $p \cdot \bar{x} \leq y$ and set $\bar{v} := u(\bar{x})$. Then

$$\sup_{x \in \mathbb{R}^\ell_{++}, p \cdot x \leq y} u(x) = \sup_{x \in C_{\bar{v}}, p \cdot x \leq y} u(x)$$

and on the right side a continuous function is maximized over a nonempty compact set.

*Remark 2.* A different application of the Kuhn-Tucker theorem is obtained if, instead of working with the above open set $\Omega \subset \mathbb{R}^\ell_+$, one uses a model where the utility function $u$ is defined and differentiable on an open set $\Omega'$ (for instance, this could be $\mathbb{R}^\ell$ itself) that *contains* $\mathbb{R}^\ell_+$. In that case optimality of $x^*$ for (1) can be expressed equivalently as optimality of $x^*$ for the following optimization problem:

$$\text{maximize } u(x) \text{ over all } x \in \Omega' \text{ with } p \cdot x \leq y \text{ and } -x_i \leq 0,\ i = 1, \ldots, \ell. \quad (3)$$

Precisely such an application was chosen in [9–11, 13], but in doing so one forms a model that no longer applies to all Cobb–Douglas or CES utility functions, because for general parameter values such functions are not differentiable at points in $\mathbb{R}^\ell_+ \backslash \mathbb{R}^\ell_{++}$! In [9, p. 131 ff.], [10, p. 50 ff.], [11, Theorem 22.1, Example 22.1] and [14, Sect. 2.2] (the latter reference discusses this for profit maximization) this has led to imprecise or incorrect formulations of their necessary first order optimality conditions and, in the case of [9–11], to an incorrect application to the Cobb–Douglas instance $(i)$.[1] The appendix to this paper provides more details to support this claim. Apart from this shortfall, the derivation of these optimality conditions is standard and can be found in [10, 11]: by applying the Kuhn-Tucker theorem [13] to (3), which has $\ell+1$ inequality constraints, one now obtains as a first-order necessary condition for optimality

$$\frac{\partial u(x^*)}{\partial x_i} \leq \lambda p_i, \text{ with equality if } x_i^* > 0,\ i = 1, \ldots, \ell$$

instead of (2) and this holds for every optimal $x^*$ in $\Omega'$, a set that now includes the boundary of $\mathbb{R}^\ell_+$. For utility functions that fit into this model, *but only for those*, this formulation gives meaningful and somewhat sharper results. These remarks do not affect [8]; however, the optimality results in that reference presuppose that the optimal solution belongs to $\mathbb{R}^\ell_{++}$, as is exemplified by the application to the Cobb–Douglas instance in Exercise 1.20 of [8]. That is not enough to allow a treatment of the corner point solution instances $(iii)$–$(iv)$. It reflects a common shortcoming of the above references: except for [14, p. 57], none would seem to treat corner point solutions analytically.

From part $(d)$ of Theorem 1 it is evident that stringent quasiconcavity can help to solve the optimization problem (1), but it should be kept in mind that this is very much due to the hypothesis that the utility function $u$ is strictly increasing: property $(S)$ ignores bundles at the lowest utility level, which is $u(0)$. For operational use I shall state a simple sufficient condition for stringent quasiconcavity. To prepare for it, I observe that the properties of $u$ in Theorem 1 cause the range of $u$ to be an

[1] Notwithstanding its general Kuhn-Tucker Theorem 1.D.3, reference [13] considers utility maximization only for utility functions defined on all of $\mathbb{R}^\ell$ (see its pp. 134–135); on pp. 223–224 this has resulted in an ad hoc solution of instance $(i)$.

interval, namely $[u(0), u_\infty)$, where $u_\infty := \sup_{x\in\mathbb{R}^\ell_+} u(x)$ (this notation allows for the possibility that $u_\infty$ equals $+\infty$). To see this, define $\psi(t) := u(t,t,\dots,t)$ for $t \geq 0$; this is a continuous, strictly increasing function on $\mathbb{R}_+$. Clearly, $u_\infty = \sup_{t\geq 0}\psi(t)$ and the supremum cannot be attained. Because for any $x \in \mathbb{R}^\ell_+$ the value $u(x)$ lies between $\psi(0)$ and $\psi(t)$ for some $t > 0$ sufficiently large (this follows by strict monotonicity and continuity of $u$), the intermediate value theorem can be invoked to finish the argument.

**Proposition 1.** *If $\{x \in \mathbb{R}^\ell_+ : u(x) > u(0)\}$ is convex and if there exists a strictly increasing function $h : (u(0), u_\infty) \to \mathbb{R}$ such that the composition mapping $x \mapsto h(u(x))$ is strictly quasiconcave on $\{x \in \mathbb{R}^\ell_+ : u(x) > u(0)\}$, then $u$ is stringently quasiconcave.*

*Proof.* By a previous remark, it is enough to prove that $u$ is quasiconcave on $\mathbb{R}^\ell_+$ and strictly quasiconcave on $\{x \in \mathbb{R}^\ell_+ : u(x) > u(0)\}$. Let $\alpha \in \mathbb{R}$. If $\alpha \leq u(0)$, then $\{u \geq \alpha\} := \{x \in \mathbb{R}^\ell_+ : u(x) \geq \alpha\} = \mathbb{R}^\ell_+$. If $\alpha > u(0)$, then for any $x, x' \in \{u \geq \alpha\} \subset \{u > u(0)\}$, $x \neq x'$, and for any $t \in (0,1)$ the given strict quasiconcavity property implies $h(u(tx + (1-t)x')) > \min(h(u(x)), h(u(x'))) \geq h(\alpha)$. So $u(tx + (1-t)x') > \alpha$ by strict monotonicity of $h$. For the set $\{u \geq \alpha\}$ I conclude that it equals $\mathbb{R}^\ell_+$ for every $\alpha \leq u(0)$ and that it is strictly convex for every $\alpha > u(0)$. Hence, $u$ is certainly quasiconcave on $\mathbb{R}^\ell_+$ and the desired strict quasiconcavity of $u$ on $\{u > u(0)\}$ also follows from the previous conclusion. □

*Example 1.* $(i)$ A utility function $u : \mathbb{R}^\ell_+ \to \mathbb{R}_+$ of Cobb–Douglas type is given by $u(x) := \Pi_{i=1}^\ell x_i^{\alpha_i}$; here all $\alpha_i > 0$. This function is not strictly quasiconcave on $\mathbb{R}^\ell_+$ (observe that $u(x_1, 0, 0, \dots, 0) = 0$). However, it is stringently quasiconcave and I use Proposition 1 to show this. Observe first that the set $\{x \in \mathbb{R}^\ell_+ : u(x) > u(0)\} = \mathbb{R}^\ell_{++}$ is convex. On $(u(0), u_\infty) = (0, +\infty)$ I choose $h(t) := \log(t)$; this is a strictly increasing function, which gives $h(u(x)) = \sum_{i=1}^\ell \alpha_i \log(x_i)$ on $\mathbb{R}^\ell_{++}$. Because each function $x_i \mapsto \log(x_i)$ is strictly concave on $\mathbb{R}_{++}$, it easily follows that the function $x \mapsto h(u(x))$ is strictly concave on $\mathbb{R}^\ell_{++}$. Therefore, it follows from Proposition 1 that $u$ is stringently quasiconcave on $\mathbb{R}^\ell_+$.

$(ii)$ A utility function of CES type is given by $u(x) := (\sum_{i=1}^\ell x_i^\rho)^{1/\rho}$ with either $(iia)$ $0 < \rho < 1$, $(iib)$ $\rho < 0$ or $(iic)$ $\rho > 1$. In case $(iic)$ $u$ is not stringently quasiconcave, as is easy to see. In case $(iia)$ the function $u$ is defined on $\mathbb{R}^\ell_+$, as usual, but in case $(iib)$ its domain of definition has to be $\mathbb{R}^\ell_{++}$, because of the negative exponent $\rho$. See also Remark 1$(ii)$ below. I observe that $u$ is strictly increasing in both cases $(iia)$ and $(iib)$ (and in case $(iic)$ as well).

$(iia)$ I will show that in case $(iia)$ $u$ is stringently quasiconcave and even strictly quasiconcave on $\mathbb{R}^\ell_+$. First, I observe that the set $\{x \in \mathbb{R}^\ell_+ : u(x) > u(0)\} = \mathbb{R}^\ell_+\backslash\{0\}$ is convex (because this only leaves out the origin, the argument can immediately be adapted to imply that $u$ is even strictly quasiconcave on $\mathbb{R}^\ell_+$). On $(u(0), u_\infty) = (0, +\infty)$ I choose $h(t) := t^\rho$, a strictly increasing function;

then $x \mapsto h(u(x)) = \sum_{i=1}^{\ell} x_i^{\rho}$ is easily seen to be strictly concave on $\mathbb{R}_+^{\ell}$, because each function $x_i \mapsto x_i^{\rho}$ is strictly concave on $\mathbb{R}_+$. So the conditions in Proposition 1 certainly hold; hence, $u$ is stringently quasiconcave on $\mathbb{R}_+^{\ell}$.

$(iib)$ As observed above, the domain of definition is $\mathbb{R}_{++}^{\ell}$ in this case. In particular, $u(0)$ is not well-defined, so stringent quasiconcavity loses its meaning (but see Remark 1$(ii)$ below for an adaptation). However, by analogy to Proposition 1, I can still show that $u$ is strictly quasiconcave on $\mathbb{R}_{++}^{\ell}$. Indeed, in this case $h(t) := -t^{\rho}$ is strictly increasing on $(0, +\infty)$ and $x \mapsto h(u(x)) = -\sum_{i=1}^{\ell} x_i^{\rho}$ is strictly concave on $\mathbb{R}_{++}^{\ell}$, because each $x_i \mapsto -x_i^{\rho}$ is strictly concave on $\mathbb{R}_+$ by $\rho < 0$. It now follows easily that $u$ is strictly quasiconcave on $\mathbb{R}_{++}^{\ell}$.

$(iii)$ Let $u : \mathbb{R}_+^2 \to \mathbb{R}$ be given by $u(x_1, x_2) = x_1^2(x_2 + 1)$, as in my instance $(iv)$. This function is not strictly quasiconcave on $\mathbb{R}_+^2$ (note that $u(0, x_2) = 0$ for all $x_2 \geq 0$). However, $u$ is stringently quasiconcave. To prove this, I observe first that the set $\{x \in \mathbb{R}_+^2 : u(x) > u(0)\} = \{(x_1, x_2) : x_1 > 0, x_2 \geq 0\}$ is convex. On $(u(0), u_{\infty}) = (0, +\infty)$ I choose $h(t) := \log(t)$, a strictly increasing function; then $x \mapsto h(u(x)) = 2\log(x_1) + \log(x_2 + 1)$ is evidently strictly concave on $\{(x_1, x_2) : x_1 > 0, x_2 \geq 0\}$ (repeat the reasoning in instance $(i)$). Hence, $u$ is stringently quasiconcave by Proposition 1.

The above example will be used for applications of Theorem 1 to the instances $(i)$, $(iia)$, $(iib)$ and $(iv)$. It is well-known that more can be said, although this is not of importance for my applications. For instance, for $\sum_i \alpha_i \leq 1$ the Cobb–Douglas utility function in part $(i)$ of Example 1 is concave on $\mathbb{R}_+^{\ell}$ and in part $(iia)$ the same is true for the CES utility function. Results from [5] or [12] can be invoked to derive such additional properties: see Remark 4 in [5, p. 123] or see Theorem 2.5.3 and its Examples 5, 6 in [12].

*Remark 3.* In [2] Aliprantis, Brown and Burkinshaw consider utility maximization for a pure exchange consumer, but their analysis extends effortlessly to that for an ordinary consumer. For $\ell = 2$ they correctly and completely solve instances $(i)$, $(iia)$ and $(iv)$ in a coherent way, based on using existence and necessary first order optimality conditions, similar to parts $(a)$ and $(b)$ of Theorem 1, aided by considerations involving strict quasiconcavity of $u$ on $\mathbb{R}_{++}^2$: see the solutions to problems 1.2.1, 1.3.2 and 1.3.4 in [1, pp. 25–26, pp. 34–35]. Instance $(iii)$ is not treated in [2] and instance $(v)$ is solved in an ad hoc fashion; the former would not seem to be out of reach of the general method presented in [2], but the latter would seem to be. Thus, in the area of utility maximization the book [2], although not devoted to general microeconomic theory as such, managed to reach further than the references mentioned above and it did so flawlessly.

For purposes going beyond mere computations, Definition 1.3.4 of [2] defines a *neoclassical preference* to be a continuous preference relation on $\mathbb{R}_+^{\ell}$ that has certain monotonicity properties. In terms of the representing utility function of such a preference relation, which exists by [10, Proposition 3.C.1], this definition comes down to the following two possibilities: either $(NC_1)$ $u$ is strongly increasing and strictly quasiconcave on $\mathbb{R}^\ell_+$ or $(NC_2)$ strongly increasing and strictly quasiconcave on $\mathbb{R}^\ell_{++}$ and $u(x) > u(x')$ for every $x \in \mathbb{R}^\ell_{++}$ and $x' \in \mathbb{R}^\ell_+ \backslash \mathbb{R}^\ell_{++}$. Here I use standard terminology from [8, 10] – the one used in [2] is somewhat different. Observe that CES utility functions satisfy $(NC_1)$ but not $(NC_2)$ and that Cobb–Douglas utility functions satisfy $(NC_2)$ but not $(NC_1)$; thus, the above definition is inherently two-pronged. I observe that if possibility $(NC_2)$ obtains, then one has actually the following special property, which seems not to have been stated explicitly in [2]: everywhere on the boundary of $\mathbb{R}^\ell_+$ the function $u$ is equal to $u(0)$ (to see this, it suffices to compare any point on the boundary with $(t, t, \dots, t)$ for sufficiently small $t > 0$ and use continuity and strict monotonicity of $u$ to finish the argument). I claim that both possibilities $(NC_1)$ and $(NC_2)$, and therefore the definition of a neoclassical preference relation as a whole, are subsumed by the more general notion of a strictly increasing utility function that is stringently quasiconcave (incidentally, this continues to hold if in $(NC_1)$ and $(NC_2)$ "strongly increasing" is replaced by the less demanding "strictly increasing"). For possibility $(NC_1)$ this is immediately obvious. As for $(NC_2)$, by continuity of $u$ and strict quasiconcavity of $u$ on $\mathbb{R}^\ell_{++}$ it easily follows that $u$ is quasiconcave on $\mathbb{R}^\ell_+$, which is the closure of $\mathbb{R}^\ell_{++}$. To show property $(S)$, recall from the discussion of $(NC_2)$ in the preceding lines that if $x, x' \in \mathbb{R}^\ell_+$, $x \neq x'$, have $u(x) = u(x') > u(0)$ then this implies $x, x' \in \mathbb{R}^\ell_{++}$. Therefore, the fact that possibility $(NC_2)$ results in strict quasiconcavity of $u$ on $\mathbb{R}^\ell_{++}$ implies $u(\frac{1}{2}x + \frac{1}{2}x') > u(x) = u(x')$. As already said, this shows that utility functions that correspond to a neoclassical preference are stringently quasiconcave. The converse is not true, even when the neoclassical preference definition is adapted so as to encompass strictly increasing utility functions: for instance, in Example 1$(iii)$ it was demonstrated that in instance $(iv)$ the utility function $u$ is stringently quasiconcave, but not strictly quasiconcave on $\mathbb{R}^2_+$; hence, for this $u$ possibility $(NC_1)$ is out of the question. On the other hand, $(NC_2)$ is also impossible because of $u(1,0) \not< u(\frac{1}{2}, \frac{1}{2})$.

## 3 Testing for Operational Usefulness

Here I shall discuss the use of Theorem 1 as a means to meet my criterion of operational usefulness for utility maximization; similar applications to expenditure minimization are also possible. So my task is to derive complete solutions for the Marshallian demand function, using Theorem 1 in a coherent way, for each of the following utility functions:

$(i)$ $u(x) = \Pi_{i=1}^{\ell} x_i^{\alpha_i}$ with all $\alpha_i > 0$,

$(ii)$ $u(x) = \left(\sum_{i=1}^{\ell} a_i x_i^{\rho}\right)^{1/\rho}$ with all $a_i > 0$ and $(a)$ $\rho \in (0,1)$, $(b)$ $\rho < 0$ or $(c)$ $\rho > 1$,

$(iii)$ $u(x) = \sum_{i=1}^{\ell} a_i x_i$, with all $a_i > 0$,

$(iv)$ $u(x_1, x_2) = x_1^2(x_2 + 1)$,

$(v)$ $u(x) = \min_{1 \le i \le \ell} b_i x_i$, with all $b_i > 0$.

These five functions are continuous and strictly increasing. In instance $(v)$ I suppose $\ell \geq 2$, so as to avoid overlap with instance $(iii)$. My solution method, which is based on familiar reasoning in optimization theory [3], goes as follows. The starting point is that Theorem 1$(a)$ guarantees that an optimal solution of (1) exists and is budget-balanced. It is useful to introduce the following term: an *optimality candidate* is a vector $x^* \in B_0 \cap \Omega$ that satisfies the first order necessary optimality condition (2). Then it follows from parts $(a)$–$(b)$ of Theorem 1 that the optimal solution of (1) must be an optimality candidate, provided that it belongs to $\Omega$. Subsequently, if $u$ happens to be quasiconcave then part $(c)$ applies and all optimality candidates (if there are any) are indeed optimal solutions. If in addition $u$ has property $(S)$, then the solution is of course complete (but only then, for I wish to determine *all* globally optimal solutions of (1) and in the previous steps the so-called *remainder set* $B_0 \backslash \Omega$ has not been inspected). In sum, if the utility function is stringently quasiconcave, then an optimality candidate, when found, is immediately known to be the unique optimal solution of (1). If there does not exist an optimality candidate or if $u$ is not quasiconcave or fails to possess property $(S)$, a careful look at the values that $u$ attains on the remainder set $B_0 \backslash \Omega$ is needed, and these values should be compared with the maximum value of all the optimality candidates already found (if any). For $\ell = 2$ the latter is easy, but for $\ell > 2$ it can be somewhat of a challenge: see my solution of instances $(iic)$, $(iii)$ and $(v)$.

It must be mentioned that the alternative optimality conditions that I mentioned in Remark 2 can also be used, but only for instances such as $(iii)$ and $(iv)$, where the utility functions are differentiable on all of $\mathbb{R}^\ell$. This gives $\ell + 1$ multipliers and it is well-known from nonlinear programming [3] that one should then work with the set of active indices of $x^* \in B_0$ (i.e., those indices $i$ for which $x_i^* = 0$). A good example of this is presented in Sect. 4.3 of [14] for a linear technology (the derivation there is nonrigorous because it is exclusively based on using first-order necessary conditions). Another observation is that ad hoc methods to solve specific problems abound. For example, it is well-known that instance $(i)$ can be solved by trivially eliminating the boundary of $\mathbb{R}^\ell_+$, after which one can apply the logarithmic transformation. However, such a transformation is already contained in Example 1$(i)$, where it is part of a fairly systematic solution method.

*Solution of instance (i)* To solve problem (1) for the utility function $u(x) := \Pi_{i=1}^\ell x_i^{\alpha_i}$ on $\mathbb{R}^\ell_+$ with all $\alpha_i > 0$, I choose $\Omega = \mathbb{R}^\ell_{++}$. From Example 1$(i)$ I already know that $u$ is stringently quasiconcave. So if I can find $x^* \in B_0 \cap \Omega$ that satisfies (2), then it must be the unique optimal solution by Theorem 1$(d)$. In search of such $x^*$, I combine (2) with $p \cdot x^* = y$ and verify concretely that the $x^*$ found belongs to $\Omega$. This is a simple algebraic task (incidentally, note that the possibility $\lambda = 0$ in (2) leads to $x^* \notin \Omega$). It yields $x_i^* = \alpha_i y/(\alpha p_i)$, $i = 1, \ldots, \ell$, where $\alpha := \sum_{j=1}^\ell \alpha_j$. This outcome happens to be strictly positive, so the unique optimal solution has been found.

*Solution of instance (iia)* To solve problem (1) for the utility function $u(x) := \left(\sum_{i=1}^\ell x_i^\rho\right)^{1/\rho}$ on $\mathbb{R}^\ell_+$ with $\rho \in (0, 1)$, I choose $\Omega = \mathbb{R}^\ell_{++}$. From Example 1$(iia)$ I know that $u$ is stringently quasiconcave.

*Step 1: special case $a_i = 1$ for all $i$.* If I can find $x^* \in B_0 \cap \Omega$ that satisfies (2), then it must be the unique optimal solution by Theorem 1$(d)$. To find such $x^*$, I combine (2) with $p \cdot x^* = y$ and then check that their solution belongs to $\Omega$ (the possibility $\lambda = 0$ in (2) can be excluded, because it leads to nonsensical expressions). This is a simple algebraic task, which gives the strictly positive expression $x_i^* = p_i^{r-1} y/(p_1^r + \cdots + p_\ell^r)$, $i = 1, \ldots, \ell$, where $r := \rho/(\rho - 1)$. So the unique optimal solution has been determined.

*Step 2: general case.* I absorb the coefficients into the variables by using the variables $\xi_i := a_i^{1/\rho} x_i$. Then the utility maximization problem for $u(x) := \left(\sum_{i=1}^{\ell} a_i x_i^\rho\right)^{1/\rho} = \left(\sum_{i=1}^{\ell} \xi_i^\rho\right)^{1/\rho}$ can be rewritten as the one already solved in step 1, provided that the prices are adapted by setting $q_i := a_i^{-1/\rho} p_i$. So by step 1 the unique optimal solution is $\xi_i^* = q_i^{r-1} y/(q_1^r + \cdots + q_\ell^r)$, i.e., $x_i^* = a_i^{1-r} p_i^{r-1} y/(a_1^{1-r} p_1^r + \cdots + a_\ell^{1-r} p_\ell^r)$.

*Solution of instance (iib)* To solve problem (1) for the utility function $u(x) := \left(\sum_{i=1}^{\ell} x_i^\rho\right)^{1/\rho}$ on $\mathbb{R}_{++}^\ell$ with $\rho < 0$, I choose $\Omega = \mathbb{R}_{++}^\ell$ and appeal to Remark 1$(ii)$. Because of Example 1$(iib)$, the solution method is almost exactly the same as the one presented above for case $(iia)$: again it follows that any $x^* \in B_0 \cap \Omega$ that satisfies (2) must be the unique optimal solution and again this leads to the same expression for the optimal solution $x_i^*$ as in case $(iia)$.

*Solution of instance (iic)* To solve problem (1) for the utility function $u(x) := \left(\sum_{i=1}^{\ell} x_i^\rho\right)^{1/\rho}$ on $\mathbb{R}_+^\ell$ with $\rho > 1$, I choose $\Omega = \mathbb{R}_{++}^\ell$. In contrast to case $(iia)$, the function $u$ is not stringently quasiconcave and this explains why I follow a completely different line of attack. It concentrates on the remainder set $B_0 \cap (\mathbb{R}_+^\ell \backslash \mathbb{R}_{++}^\ell)$ and uses parts $(a)$ and $(b)$ of Theorem 1.

*Step 1: special case $a_i = 1$ for all $i$.* By Theorem 1$(a)$ an optimal solution of (1) for $u(x) := \left(\sum_{i=1}^{\ell} x_i^\rho\right)^{1/\rho}$ exists and it belongs to $B_0$. I claim that any such optimal solution must have precisely $\ell - 1$ coordinates equal to 0. This can be proved by means of a contradiction argument. For suppose that an optimal solution $x^*$ of (1) had precisely $k > 1$ nonzero coordinates; without loss of generality I can suppose that these are the first $k$ coordinates $x_1^*, \ldots, x_k^*$ (or else I could relabel the coordinates). The given optimality of $x^*$ implies a fortiori that $z^* := (x_1^*, \ldots, x_k^*)$, the vector composed of the first $k$ coordinates of $x_k^*$, is the optimal solution of the following problem: maximize $\tilde{u}(z) := \left(\sum_{i=1}^{k} z_i^\rho\right)^{1/\rho}$ over all $z \in \mathbb{R}_+^k$ such that $\tilde{p} \cdot z \leq y$. Here $\tilde{p} := (p_1, \ldots, p_k)$. Observe that $z^*$ belongs to $\mathbb{R}_{++}^k$ by definition of $k$. So I can apply Theorem 1$(b)$ to the aforementioned maximization problem, which is a version of (1) with dimension $k$. From the theorem it follows that there exists $\lambda \geq 0$ such that $\nabla \tilde{u}(z^*) = \lambda \tilde{p}$. Writing this out, the same algebra as in case $(iia)$ gives $z_i^* = p_i^{r-1} y/(p_1^r + \cdots + p_k^r)$, $i = 1, \ldots, k$. Because $z_i^* = x_i^*$ for all $i \leq k$ and $x_i^* = 0$ for all $i > k$, it would follow with a little algebra that

$x^* = (x_1^*, \dots, x_k^*, 0, \dots, 0)$ satisfies $u(x^*) = y/\left(\sum_{i=1}^k p_i^r\right)^{1/r}$. But the present case has $r := \rho/(\rho - 1) > 0$, so $u(y/p_1) > u(x^*)$ would hold by $k \geq 2$ and this contradiction proves my claim. It follows easily that the optimal solution, known to exist, must be one of the corner points $x^i := (0, \dots, 0, y/p_i, 0, \dots, 0)$, $i = 1, \dots, \ell$, of $B_0$. It is now easy to conclude that any corner point $x^j$ whose index $j$ is such that it corresponds to the lowest price, i.e., $p_j = \min_{1 \leq i \leq \ell} p_i$, is optimal (so multiple optimal solutions can occur when two or more commodities share the lowest price).

*Step 2: general case.* The same absorption trick as used in case $(ii a)$, i.e., setting $\xi_i := a_i^{1/\rho} x_i$ and $q_i := a_i^{-1/\rho} p_i$, gives precisely the same optimal solution(s) as obtained in step 1, i.e., the corner point(s) of $B_0$ which correspond to the lowest value(s) $a_i^{-1/\rho} p_i$.

*Solution of instance (iii)* To solve problem (1) for the linear utility function $u(x) := \sum_{i=1}^{\ell} a_i x_i$, with all $a_i > 0$, I choose $\Omega = \mathbb{R}^{\ell}_{++}$. The function $u$ is clearly not stringently quasiconcave, but it is so simple that a solution, essentially based on parts $(a)$ and $(b)$ of Theorem 1, can be provided in each of the following two cases. Case 1: $a := (a_1, \dots, a_\ell)$ is a scalar multiple of $p$, say $a = \mu p$ for some $\mu \in \mathbb{R}$, and then $\mu > 0$ of course. Case 2: $a$ is not a scalar multiple of $p$.

*Case 1:* By Theorem 1$(b)$, for any optimal $x^* \in \Omega := \mathbb{R}^{\ell}_{++}$ there exists $\lambda \geq 0$ such that $a = \lambda p$. In the present case this was already true (take $\lambda = \mu > 0$), so every $x^* \in B_0 \cap \Omega$ is an optimality candidate. Next, $u$ is quasiconcave, so every optimality candidate is also an optimal solution. Again this offers no news, because all optimality candidates $x^*$ satisfy $a \cdot x^* = \mu p \cdot x^* = \mu y$ and because $a \cdot x = \mu p \cdot x \leq \mu y$ for all $x \in B$. It remains to inspect the remainder set $B_0 \backslash \Omega$. Here that is the set of all $x^*$ in $B_0$ with at least one coordinate equal to zero. The value $u(x^*)$ of any $x^* \in B_0 \backslash \Omega$ is $a \cdot x^* = \mu p \cdot x^* = \mu y$, which is the same value as found before. Hence, I conclude that in case 1 the set of all optimal solutions is $B_0$, that is to say the union of $B_0 \cap \Omega$ and $B_0 \backslash \Omega$.

*Case 2:* This time the necessary condition $a = \lambda p$ is incompatible with the present situation. So there are no optimality candidates at all, which means that the optimal solution, known to exist by Theorem 1$(a)$, must belong to the remainder set $B_0 \backslash \Omega$. Now for every $x$ in $B$ I have $a \cdot x = \sum_i a_i p_i^{-1} p_i x_i \leq \alpha p \cdot x \leq \alpha y$, where $\alpha := \max_i a_i/p_i$. Let $I$ be the set of those indices $i$ for which $a_i/p_i = \alpha$. Then $I$ is nonempty and it is a strict subset of $\{1, \dots, \ell\}$ (or else I would find $a = \alpha p$, which cannot be true in the present case). Any corner point $x^i := (0, \dots, 0, y/p_i, 0, \dots, 0)$, $i \in I$, belongs to $B_0$ and achieves $a \cdot x^i = \alpha y$. So I conclude $\sup_{x \in B} a \cdot x = \alpha y$. I claim that the set of optimal solutions is the intersection $B_0 \cap \bigcap_{i \notin I} \{x \in \mathbb{R}^{\ell}_{+} : x_i = 0\}$, which is a face of $B_0$. As for one inclusion, let $x^*$ belong to this intersection. Then $a \cdot x^* = \sum_{i \in I} a_i x_i^* = \sum_{i \in I} a_i p_i^{-1} p_i x_i^* = \alpha p \cdot x^* = \alpha y$, so $x^*$ is optimal. Conversely, let $x^*$ be optimal. Then Theorem 1$(a)$ implies $x^* \in B_0$ and $\alpha y = a \cdot x^*$. Hence, $\alpha y = \alpha \sum_{i \in I} p_i x_i^* + \sum_{i \notin I} a_i x_i^*$. By $p \cdot x^* = y$ this implies $\sum_{i \notin I} (a_i - \alpha p_i) x_i^* = 0$, so it follows that $x_i^* = 0$ for every $i \notin I$ (note that $a_i - \alpha p_i < 0$ for each

$i \notin I$). This proves the desired characterization of the set of optimal solutions in case 2.

*Solution of instance (iv)* To solve problem (1) for the utility function $u(x_1, x_2) := x_1^2(x_2 + 1)$ on $\mathbb{R}^2_+$, I choose $\Omega = \mathbb{R}^2_{++}$. By Example 1(*iii*) $u$ is stringently quasiconcave. So if I can find $x^* \in B_0 \cap \Omega$ which satisfies (2), then it must be the unique optimal solution by Theorem 1($d$). I solve $p \cdot x^* = y$ and (2), the latter amounting to $x_2^* = p_1(2p_2)^{-1}x_1^* - 1$, and then I pick the solution, if any, that belongs to $\Omega$. From the former two equations I also obtain that $\lambda = 0$ if and only if $(x_1^*, x_2^*) = (0, y/p_2)$, a vector which is not in $\Omega$. So I can proceed with $\lambda > 0$. I solve the two equations and this gives $x_1^* = \frac{2}{3}(y + p_2)/p_1$, $x_2^* = \frac{1}{3}(y - 2p_2)/p_2$. Because $x^* \in \Omega$ is needed for $x^*$ to be an optimality candidate, I distinguish between the following two cases (note that $y > 2p_2$ is equivalent to $x^* \in \Omega$):

*Case 1:* $y > 2p_2$. Then $x^* = (\frac{2}{3}(y + p_2)/p_1, \frac{1}{3}(y - 2p_2)/p_2)$ belongs to $\Omega$. Hence, it is an optimality candidate and by the previous argument it must also be the unique optimal optimal solution.

*Case 2:* $y \le 2p_2$. In this case there is not any optimality candidate. Therefore, I know that the best $u$-value over the remainder set $B_0 \backslash \Omega$ is the optimal value. The set $B_0 \backslash \Omega$ contains only two vectors, the corner points $(y/p_1, 0)$ and $(0, y/p_2)$. Of these, the former one gives the highest $u$-value. Combining cases 1 and 2, I conclude that the optimal solution is given by

$$(x_1^*, x_2^*) = \begin{cases} (\frac{2}{3}(y + p_2)/p_1, \frac{1}{3}(y - 2p_2)/p_2) & \text{if } y > 2p_2, \\ (y/p_1, 0) & \text{if } y \le 2p_2. \end{cases}$$

This is only partly a corner solution; graphical illustrations can help greatly to understand why this is so.

*Solution of instance (v)* To solve problem (1) for the utility function $u(x) = \min_{1 \le i \le \ell} b_i x_i$ on $\mathbb{R}^\ell_+$ with all $b_i > 0$, I choose $\Omega$ to be the open set of all $x \in \mathbb{R}^\ell_{++}$ such that $b_i x_i \ne b_j x_j$ for any $i \ne j$. Then $u$ is locally of the form $u(x) = b_k x_k$ near each point $x^*$ of $\Omega$ (the index $k$ being unique to $x^*$), so there are no optimality candidates at all, because in such $x^*$ condition (2) amounts to $b_k e_k = \lambda p$, which does not have a solution (recall that $\ell \ge 2$). Here $e_k$ again denotes the $k$-th unit vector. It follows that the optimal solution, known to exist, must belong to the remainder set $B_0 \backslash \Omega$. For any $x$ in $B_0 \backslash \Omega$ either of the following two cases can occur: case 1: $x \in \mathbb{R}^\ell_{++}$ and there is a tie in the form of an equality $b_i x_i = b_j x_j$ for some $i \ne j$ or case 2: some coordinate of $x$ is equal to zero. In case 2 $u(x)$ equals zero, which means the certain sub-optimality of $x$. So I concentrate on those vectors in $B_0 \cap \mathbb{R}^\ell_{++}$ that have ties. For any $x$ in $B$ I have $x_i \ge u(x)/b_i$ for all $i$, giving $y \ge \sum_i p_i x_i \ge u(x) \sum_i p_i/b_i$, i.e., $u(x) \le y/\beta$, with $\beta := \sum_i p_i/b_i > 0$. In fact, the same reasoning shows that $u(x) < y/\beta$ whenever the set of all indices $i$ for which $b_i x_i > u(x)$ is nonempty. So $u(x^*) = y/\beta$ for $x^* \in B$, which amounts to $x^*$ being optimal in view of the previous lines, requires $b_i x_i^* = b_j x_j^* = u(x^*)$ for all $i$ and $j$. Clearly, there is precisely one such $x^*$ and it is given by $x_i^* = b_i^{-1}y/\beta$, $i = 1, \ldots, n$; so this is the optimal solution.

**Acknowledgements** I am indebted to Jean-Pierre Crouzeix (Université Blaise Pascal, France) for the reported observations in connection with reference [5] and to the late Roko Aliprantis for calling my attention to his books [1, 2] and for helpful remarks. I thank Alexander Gnedin (University of Utrecht) for helpful criticism and encouragement. I also thank an anonymous referee for several very helpful remarks.

## Appendix

This appendix details the findings that I summarized in Remark 2. I discuss what I found in the textbooks [8, 10, 11, 13], all of which make explicit claims about being rigorous and precise, and also in the two less formally oriented textbooks [9, 14] which were added in a second round. Reference [7] is also reviewed briefly, at the request of one of the editors. The intermediate textbook [4], which I cite for other reasons, is no part of these comparisons.

*Example 1. (Mas-Colell-Whinston-Green [10])* In *Microeconomic Theory* [10, p. 50] the utility function $u$ is defined on the consumption set $\mathbb{R}^\ell_+$. On p. 53, which states first-order necessary conditions for optimality in problem (1), $u$ is required to be continuously differentiable in the following way: *"If $u(\cdot)$ is continuously differentiable, an optimal consumption bundle ... can be characterized ..."*. I quote this in full, because what is meant by differentiability of functions on closed sets can be a very confusing matter (for instance, some authors will call a function differentiable on $\mathbb{R}^\ell_+$ if and only if it is differentiable in every interior point of $\mathbb{R}^\ell_+$, i.e., if it is differentiable on $\mathbb{R}^\ell_{++}$; see [13, p. 81]). In Example 3.D.1O [10, p. 56] the first-order necessary conditions are applied to the Cobb–Douglas instance $u(x_1, x_2) = kx_1^\alpha x_2^{1-\alpha}$, with $0 < \alpha < 1$. This function does not have partial derivatives in points with either $x_1 = 0$ or $x_2 = 0$. In view of this application, it cannot be that the necessary conditions on p. 53 should contain a partial derivative $\frac{\partial u(x^*)}{\partial x_i}$ for any $x^*$ on the boundary of $\mathbb{R}^\ell_+$. Yet this is so: the first-order condition (3.D.1) on p. 53 contains the following explicit clause *"with equality if $x_i^* > 0$"*, which would make no sense if boundary points $x^*$ were excluded a priori. In fact, on p. 54 one finds: *"Fig. 3.D.4(b) depicts the first-order condition ... when the consumer's optimal bundle $x^*$ lies on the boundary of the consumption set."* The trouble is that the proof on p. 53 of [10] relies on the Kuhn-Tucker Theorem M.K.2 from p. 959. But the latter result is formulated for functions whose domain is all of $\mathbb{R}^\ell$. See Remark 2 for further explanation. To those readers who might object that, after all, in Example 3.D.1 boundary points are automatically sub-optimal and need not really be subjected to (3.D.1), the response must be that, mathematically/logically speaking, such an ad hoc readjustment is beside the point: necessary conditions for optimality should apply to *all* points for which they have been formulated. Here the result and the function were simply not made for each other. There is also another misplaced application of first order necessary conditions in [10]. Example 3.D.1 on p. 56 treats the Cobb–Douglas family in a well-known ad hoc fashion by the logarithmic transformation. However, the transformed utility function is once more *hors concours* for the first-order condition (3.D.1), which is formulated for a *real-valued* function that is defined on all of $\mathbb{R}^\ell_+$. Writing *"Since* $\log 0 = -\infty$ ..." on [10, p. 55] cannot alter this fact.

*Example 2. (Simon–Blume [11])* Example 22.1 in *Mathematics for Economists* [11] treats utility maximization for the Cobb–Douglas instance with $\ell = 2$. In Example 22.1 one finds the following: *"We use Theorems 22.1 and 22.2 to compute the demand function ... for the Cobb–Douglas utility function* $U(x_1, x_2) = x_1^a x_2^b$.*"* Here $a, b > 0$ and $a + b = 1$. However, Theorems 22.1 and 22.2 have the same deficiency as observed in Example 1: they are formulated for a utility function $U : \mathbb{R}^\ell_+ \to \mathbb{R}$ that is stated to be $C^1$ (i.e., continuously differentiable). Just as on [10, p. 53], the condition *"... and equality holds if* $x_i^* > 0$*"* appears in a formula that uses partial derivatives of the utility function at the point $(x_1^*, \ldots, x_\ell^*)$. So, just as in [10] there is a mismatch: these results require derivatives to be well-defined on all of $\mathbb{R}^\ell_+$, a requirement that the above Cobb–Douglas utility function cannot meet. Compared to [10], the book by Simon and Blume offers a longer discussion of (continuous) differentiability. In fact, on p. 371 its definition is formally stated, but only for a function defined on all of $\mathbb{R}^\ell$.

*Example 3. (Takayama [13])* The approach taken in *Mathematical Economics* [13] suffers from similar shortcomings. The utility maximization problem is stated on [13, p. 135] for a utility function that appears to be defined on all of $\mathbb{R}^\ell$. The first-order necessary conditions for optimality are obtained directly from the Kuhn-Tucker type result on p. 134, which works with $\mathbb{R}^\ell$ for the domain of the functions. Upon closer inspection, it is surprising that a far better Kuhn-Tucker type result is available in [13, Theorem 1.D.3]. This result allows open domains of definition for the functions involved in the optimization problem, but the aforementioned application on pp. 134–135 does not benefit from this. The subsequent treatment of the Cobb–Douglas instance in [13, pp. 223–224] (for $\ell = 2$) is presented correctly, but with no apparent connections to the material on pp. 134–135 (see also footnote 32 on p. 231 of [13]).

*Example 4. (Jehle-Reny [8])* As already indicated in Sect. 2, the first-order optimality conditions on p. 23 ff. in *Advanced Microeconomic Theory* [8] are virtually the same as parts $(a)$–$(c)$ of Theorem 1. Nonetheless, there are serious complications with the way [8, Chapter A2] presents the underpinning notions for its optimality conditions. This is because [8] fails to distinguish (total) differentiability from partial differentiability: only a formal definition of partial differentiability is presented, namely in Definition A2.1. Consequently, Example 4 of [6, p. 69] forms a counterexample against the intuitive definition of the directional derivative on [8, p. 463], which is exclusively based on partial differentiability. This can also help to explain why no chain rule for (totally) differentiable functions has been presented in [8]; yet this rule is applied. Moreover, a formula for the total differential can be found on pp. 487–488, but neither the adjective "total" nor the term "differential" would seem to have been defined in [8]. Last but not least, as already mentioned in Sect. 2 the treatment of (1) in [8] is exclusively occupied with optimal solutions which already belong to $\mathbb{R}^\ell_{++}$. Therefore, it is unable to treat situations with corner point solutions, such as instances *(iii)*–*(iv)*. Hence its Exercise 1.20 about deriving the Marshallian demand for the Cobb–Douglas instance seems slightly incomplete.

*Example 5. (Luenberger [9])* On p. 130 of *Microeconomic Theory* [9] the standard choice of $\mathbb{R}^\ell_+$ is made for the commodity space. The first-order conditions on p. 131 and p. 133 are correctly stated for interior points of $\mathbb{R}^\ell_+$. However, at the bottom of p. 133, still in the same section and with no mention of any change of the model, the following is stated: *"We assumed above that the solution occurred at a point* $\mathbf{x}$ *with* $\mathbf{x} > \mathbf{0}$ *... If the solution occurs at a point with* $x_1 > 0$ *and* $x_2 = 0$*, the first-order condition is that* $MRS_{21} > p_1/p_2$*."* Here the notation refers to marginal rate of substitution, a notion which contains partial derivatives that would have to be meaningful on the boundary but need not be so for the Cobb–Douglas family. Yet Example 5.1 on p. 132 discusses application of the same first-order conditions to the Cobb–Douglas utility function.

*Example 6. (Varian [14]) Microeconomic Analysis* [14] contains no first-order optimality conditions that are especially made to suit the general utility maximization problem. Rather, for each separate application the author refers to the Kuhn-Tucker theorem on p. 503. However, this result is only formulated for functions with all of $\mathbb{R}^\ell$ as their domain of definition. Hence, applications to the Cobb–Douglas instance (p. 111) and the CES instance (p. 55) are out of order from a formal viewpoint (on p. 112 utility maximization for the CES family is treated by means of earlier results for the producer that are handled by duality). Of course, the use of the logarithmic transformation to the Cobb–Douglas function on p. 111 could be exploited very easily to rectify the application of the Kuhn-Tucker theorem, because the transformed optimization problem takes place on $\mathbb{R}^2_{++}$, which is an open set (see Remark 2). In contrast to all other references under review, [14, p. 57] contains a rather complete derivation (informal, because it is based only on applying first-order necessary conditions) of a corner point solution. This was already observed in Sect. 3.

*Example 7. (Henderson-Quandt [7]) Microeconomic Theory: a Mathematical Approach* [7] offers more evidence of the aforementioned confusion. The conditions for the utility function, stated in the lines following (2-1) on p. 8, fail to include a proper specification of the domains of definition, differentiability and continuity. General existence of an optimal solution seems to have been ignored. Worse, because the conditions on p. 8 also require the partial derivatives of the utility function to be *strictly* positive, the applications to the Cobb–Douglas utility function in [7, pp. 18–19] can only be explained if the main results of Sects. 2-2 and 2-6 were meant to deal with $\mathbb{R}^{\ell}_{++}$ as the domain of the utility function. However, that clashes with the corner solution depicted in Fig. 2-4(b) and the nonnegativity formulas that precede it. The lines following (2-1) also require the utility function to be regular strictly quasiconcave, which is in terms of second-order partial derivatives (see footnote 1 on p. 8); thus, some mention can be made in [7] about sufficiency in the sense of part $(c)$ of Theorem 1. However, a coherent and complete method of utility maximization is not presented and, in fact, because of the deficiencies observed above, already Sect. 2-2 (which is for $\ell = 2$) is unable to deal with any of the five instances $(i)$ to $(v)$ in a mathematically correct way.

## References


1. Aliprantis, C.D.: Problems in Equilibrium Theory. Springer, New York (1996)
2. Aliprantis, C.D., Brown, D.J., Burkinshaw, O.: Existence and Optimality of Competitive Equilibria. Springer, New York (1989)
3. Bazaraa, M.S., Sherali, H.D., Shetty, C.M.: Nonlinear Programming (2nd ed.). Wiley, New York (1993)
4. Besanko, D., Braeutigam, D.R.: Microeconomics (2nd ed.). Wiley, New York (2005)
5. Crouzeix, J.-P.: Conditions for convexity of quasiconvex functions. Math. Oper. Res. **5**, 120–125 (1980)
6. Edwards, C.H.: Advanced Calculus of Several Variables. Academic Press, New York (1973)
7. Henderson, J.M., Quandt, R.E.: Microeconomic Theory: a Mathematical Approach (3rd ed.). McGraw-Hill, New York (1980)
8. Jehle, G.A., Reny, P.J.: Advanced Microeconomic Theory (2nd ed.). Wiley, New York (2001)
9. Luenberger, D.G.: Microeconomic Theory. McGraw-Hill, New York (1995)
10. Mas-Colell, A., Whinston, M.D., Green, J.R.: Microeconomic Theory. Oxford University Press, New York (1995)
11. Simon, C.P., Blume, L.: Mathematics for Economists. Norton, New York (1993)
12. Sydsæter, K., Hammond, P., Seierstad, A., Strøm, A.: Further Mathematics for Economic Analysis. Prentice Hall, Harlow, England (2005)
13. Takayama, A.: Mathematical Economics (2nd ed.). Cambridge University Press, Cambridge (1996)
14. Varian, H.R.: Microeconomic Analysis (3rd ed.). Norton, New York (1992)

# On the Definitions of Giffen and Inferior Goods

**Pierre von Mouche and Willem Pijnappel**

**Abstract** We consider Giffen goods for the very general setting of an upper semi-continuous utility function. Two not equivalent definitions for this notions are discussed. Concerning this we show that there exists a (well-defined) non-decreasing demand function $\mathbb{R}_{++} \to \mathbb{R}$ that is nowhere strictly increasing. Also we reconsider the relation between Giffen and inferior goods.

## 1 Introduction

Consider, for $n$ goods, the utility maximisation problem under a budget restriction. For this setting, in the literature there exist essentially two ways of defining what it means that a good is Giffen. The definitions have in common that for good $i$ to be Giffen there should exist prices $\mathbf{p}_{\hat{i}} = (p_1, \ldots, p_{i-1}, p_{i+1}, \ldots, p_n)$ of the other goods and a budget $m$ such that the demand correspondence $\breve{\mathbf{x}}$, assigning to the price $p_i$ of good $i$ the set of maximisers of the utility maximisation problem for $\mathbf{p} = (p_1, \ldots, p_n) =: (p_i; \mathbf{p}_{\hat{i}})$ and $m$ has some specific property.

One definition states that good $i$ is Giffen if there exist $\mathbf{p}_{\hat{i}}, m$ and two prices $p_i, p_i'$ with $p_i' < p_i$ such that

$$\breve{x}_i(p_i') < \breve{x}_i(p_i). \tag{1}$$

P. von Mouche (✉)
Wageningen Universiteit, Hollandseweg 1, Postbus 8130, 6700 EW Wageningen, The Netherlands
e-mail: pvmouche@yahoo.fr

W. Pijnappel
Spechtstraat 72, 5932 VK Tegelen, The Netherlands
e-mail: pijnappel2@yahoo.com

W. Heijman and P. Mouche (eds.), *New Insights into the Theory of Giffen Goods*, Lecture Notes in Economics and Mathematical Systems 655,
DOI 10.1007/978-3-642-21777-7_4, © Springer-Verlag Berlin Heidelberg 2012

The other is

$$\breve{x}_i \text{ is strictly increasing on a non-degenerate interval.}^1 \tag{2}$$

For (1) and (2) to have a meaning, it is not necessary that $\breve{x}_i$ is singleton-valued in the points where it is evaluated in these definitions: just interpret the $' <'$-sign in an appropriate way.[2] It should be said that concerning the literature where singleton-valuedness is not assumed when defining Giffen goods, we only are aware of [1].[3] Especially (1) quite often is used in elementary textbooks where the existence of Giffen goods is illustrated by a graphical analysis for the case of two goods with a figure showing (two) indifference sets and two budget lines. Of course, if there exists an interval $I$ such that $\breve{x}_i \upharpoonright I$ is singleton-valued, then sufficient for (2) to hold is that there exists $p_i \in I$ in which $\breve{x}_i \upharpoonright I$ is differentiable with $\frac{d\breve{x}_i}{dp_i} > 0$.

If (2) holds, then (1) holds. But, if (1) holds, then (2) does not necessarily hold as we see from the following example:

*Example 1.* Consider for prices $p_i' < p_i$ of good $i$ the corresponding budget hyperplanes (so the budget lines if $n = 2$), say $H'$ respectively $H$. Take $\mathbf{x}' \in H' \backslash H$ and $\mathbf{x} \in H \setminus H'$ such that $x_i' < x_i$. Define the utility function $u : \mathbb{R}^n_+ \to \mathbb{R}$ by $u(\mathbf{x}) = 137$, $u(\mathbf{x}') = 137\frac{1}{3}$ and $u = 0$ elsewhere. Now $u$ is upper semi-continuous, $\breve{x}_i(a) = \{x_i'\}$ $(a \leq p_i')$, $\breve{x}_i(a) = \{x_i\}(p_i' < a \leq p_i)$, $\breve{x}_i(a) = [0, m/a](a > p_i)$. ◇

In the first instance one might think that one cannot construct such an example anymore, and thus that (1) and (2) are equivalent, in case $\breve{x}_i$ is continuous (and singleton-valued). However, in Sect. 5 we show with Theorem 2, using a deep result from the theory of real functions and a (technical) construction that also for this situation they are not equivalent. The result concerns the fact that there exists a differentiable nowhere monotone[4] function of bounded variation. The construction concerns the following: out of a given continuous function of bounded variation $f_1 : [a, b] \to \mathbb{R}$ such that $0 < f_1(p_1) < m/p_1$ $(a \leq p_1 \leq b)$, a continuous utility function $u$ is constructed with $f_1$ as demand function (for good 1) for prices $a \leq p_1 \leq b$.

The first definition, i.e. (1), is our favorite one: it is the most friendly one for a good to be Giffen. In Sect. 3 we will elaborate on it, assuming only upper semi-continuity of the utility function. For the notion of inferior good the same procedure is followed. So, in particular no assumptions are made such as maximisers are budget-balanced or Marshallian demand and Engel correspondences are singleton-valued.

---

[1]That is the interval contains infinitely many points.

[2]Indeed: for subsets $A$ and $B$ of $\mathbb{R}$, $A \leq B$ means $a \leq b$ $(a \in A, b \in B)$ and $A < B$ means $a < b$ $(a \in A, b \in B)$.

[3]But also see Exercise 2.F.13 in [5].

[4]'Nowhere monotone' means that there is no non-degenerate interval on which the function is monotone.

In Sect. 4, Theorem 1, it is shown that every Giffen good (in the sense of (1)) is inferior.[5] This theorem, obtained by quite subtle reasoning, might be stronger then all others of its kind in the literature.

## 2 Setting

Given a (utility) function $u : \mathbb{R}^n_+ \to \mathbb{R}$, (budget) $m \geq 0$ and (prices) $\mathbf{p} \in \mathbb{R}^n_{++}$, consider the (utility) maximisation problem

$$\text{maximise } u(\mathbf{x}) \text{ over all } \mathbf{x} \in \mathbb{R}^n_+ \text{ such that } \mathbf{p} \cdot \mathbf{x} \leq m. \qquad \dagger_{\mathbf{p};m}$$

If $u$ is upper semi-continuous, then each utility maximisation problem has at least one maximiser. In order to have at most one maximiser it is sufficient that $u$ is strictly quasi-concave. And sufficient for each maximiser $\mathbf{x}$ to be budget-balanced, i.e. to satisfy $\mathbf{p} \cdot \mathbf{x} = m$, is that $u$ is locally non-satiated. From now on we only make the implicit assumption that $u$ is **upper semi-continuous**.

Define the correspondence

$$\tilde{\mathbf{x}} : \mathbb{R}^n_{++} \times \mathbb{R}_+ \multimap \mathbb{R}^n$$

by: $\tilde{\mathbf{x}}(\mathbf{p}; m)$ is the set of maximisers of $\dagger_{\mathbf{p};m}$. This correspondence is proper, i.e. $\tilde{\mathbf{x}}(\mathbf{p}; m) \neq \emptyset$ for all $\mathbf{p}$ and $m$, and is called the 'Marshallian demand correspondence'. For a good $i$, let $\tilde{x}_i$ be the $i$-th component projection of $\tilde{\mathbf{x}}$, i.e. the correspondence $\tilde{x}_i : \mathbb{R}^n_{++} \times \mathbb{R}_+ \multimap \mathbb{R}$ defined by

$$\tilde{x}_i(\mathbf{p}; m) := \{x_i \mid \mathbf{x} \in \tilde{\mathbf{x}}(\mathbf{p}; m)\}.$$

For a good $i$ and fixed budget $m$ and prices $\mathbf{p}_{\hat{\imath}}$ of the other goods, define

- the 'demand correspondence' $\check{x}_i : \mathbb{R}_{++} \multimap \mathbb{R}$ by $\check{x}_i(p_i) := \tilde{x}_i(\mathbf{p}; m)$;
- the 'demand function' as the restriction $\check{x}_i \upharpoonright I : I \to \mathbb{R}$, where $I$ is the subset of prices $p_i$ for which $\dagger_{\mathbf{p};m}$ has a unique maximiser.[6]

For a good $i$ and fixed prices $\mathbf{p}$, define

- the 'Engel correspondence' $\hat{x}_i : \mathbb{R}_+ \multimap \mathbb{R}$ by $\hat{x}_i(m) := \tilde{x}_i(\mathbf{p}; m)$;
- the 'Engel function' of $i$ as the restriction $\hat{x}_i \upharpoonright J : J \to \mathbb{R}$, where $J$ is the subset of budgets $m$ for which $\tilde{\mathbf{x}}(\mathbf{p}; m)$ has a unique maximiser.

[5] Also see the contribution [7] in this book for a relation in the sense of (2).

[6] We do not discriminate between a set with one element (i.e. a singleton) and the element itself.

## 3 Definition of Giffen and Inferior Goods

The following two definitions are not only very natural but also very friendly for a utility function to be Giffen or inferior. The definitions refer to the correspondences $\check{x}_i$ and $\hat{x}_i$.

**Definition 1.** Consider an upper semi-continuous utility function $u : \mathbb{R}^n_+ \to \mathbb{R}$.

- Let $i$ be a good. Fix a budget $m$ and prices $\mathbf{p}_{\hat{\imath}}$ of the other goods.
  - Let $\mathscr{P}$ be a price section.[7]
    We say that good $i$ is 'ordinary on $\mathscr{P}$' at $(\mathbf{p}_{\hat{\imath}}; m)$ if
    $$[p_i, p_i{}' \in \mathscr{P} \text{ with } p_i{}' < p_i] \;\Rightarrow\; \check{x}_i(p_i{}') \geq \check{x}_i(p_i).$$
    We say that good $i$ is 'Giffen on $\mathscr{P}$' at $(\mathbf{p}_{\hat{\imath}}; m)$ if
    $$[p_i, p_i{}' \in \mathscr{P} \text{ with } p_i{}' < p_i] \;\Rightarrow\; \check{x}_i(p_i{}') < \check{x}_i(p_i).$$
  - We say that good $i$ is 'ordinary' at $(\mathbf{p}_{\hat{\imath}}; m)$ if it is ordinary on $\mathbb{R}_{++}$ at $(\mathbf{p}_{\hat{\imath}}; m)$.
    We say that good $i$ is 'Giffen at $(\mathbf{p}_{\hat{\imath}}; m)$' if there is a price section on which it is Giffen at $(\mathbf{p}_{\hat{\imath}}; m)$.
- We say that good $i$ is 'ordinary' if it is ordinary at all $(\mathbf{p}_{\hat{\imath}}; m)$.
  We say that good $i$ is 'Giffen' if there is $(\mathbf{p}_{\hat{\imath}}; m)$ such that good $i$ is Giffen at $(\mathbf{p}_{\hat{\imath}}; m)$.
- We say that $u$ is 'ordinary' if each good is ordinary.
  We say that $u$ is 'Giffen' if there is some good that is Giffen. ◇

*Remarks.* 1. We say 'good $i$ is Giffen on $\mathscr{P}$' if there exists $(\mathbf{p}_{\hat{\imath}}; m)$ such that good $i$ is Giffen on $\mathscr{P}$ at $(\mathbf{p}_{\hat{\imath}}; m)$.
2. It is impossible that a good is Giffen and ordinary.
3. The statement 'good $i$ is not Giffen $\Rightarrow$ good $i$ is ordinary', does not hold (indeed: take $u = 0$). But it holds in case each utility maxisation problem has a unique maximiser.[8] In this case each good (and $u$) is ordinary or Giffen.
4. It is impossible that a good is Giffen on $\mathbb{R}_{++}$ at some $(\mathbf{p}_{\hat{\imath}}; m)$ because each price section on which a good is Giffen is (due to the fixed budget) bounded.

---

[7] That is, a subset of $\mathbb{R}_{++}$ with $\#\mathscr{P} \geq 2$.

[8] Indeed: suppose good $i$ is not Giffen. This implies that for every $(\mathbf{p}_{\hat{\imath}}; m)$ and price section $\mathscr{P}$ good $i$ is not Giffen on $\mathscr{P}$ at $(\mathbf{p}_{\hat{\imath}}; m)$. In particular for every $p_i{}' < p_i$ and $(\mathbf{p}_{\hat{\imath}}; m)$, $\check{x}_i(p_i{}') < \check{x}_i(p_i)$ does not hold. So, because $\#\check{x}_i(p) = 1$ $(p > 0)$, we have for every $p_i{}' < p_i$ and $(\mathbf{p}_{\hat{\imath}}; m)$ that $\check{x}_i(p_i{}') \geq \check{x}_i(p_i)$. It follows for every $(\mathbf{p}_{\hat{\imath}}; m)$ that good $i$ is ordinary on $\mathbb{R}_{++}$ at $(\mathbf{p}_{\hat{\imath}}; m)$. Thus good $i$ is ordinary.

5. For $n = 1$ there does not exist a (upper semi-continuous) Giffen utility function.[9]
6. Example 1 gives in case $n \geq 2$, a concrete discontinuous Giffen utility function: Providing a concrete continuous Giffen utility function is a little bit puzzling, even in case $n = 2$.[10]

**Definition 2.** Consider an upper semi-continuous utility function $u : \mathbb{R}^n_+ \to \mathbb{R}$.

- Let $i$ be a good. Fix prices $\mathbf{p}$.
  - Let $\mathscr{M}$ be a budget section.[11]
    We say that good $i$ is 'normal on $\mathscr{M}$' at $\mathbf{p}$ if

$$[m, m' \in \mathscr{M} \text{ with } m' < m] \Rightarrow \hat{x}_i(m') \leq \hat{x}_i(m).$$

    We say that good $i$ is 'inferior on $\mathscr{M}$' at $\mathbf{p}$ if

$$[m, m' \in \mathscr{M} \text{ with } m' < m] \Rightarrow \hat{x}_i(m') > \hat{x}_i(m).$$

  - We say that good $i$ is 'normal at $\mathbf{p}$' if it is normal on $\mathbb{R}_+$ at $\mathbf{p}$.
    We say that good $i$ is 'inferior at $\mathbf{p}$' if there is a budget section on which it is inferior at $\mathbf{p}$.
- We say that good $i$ is 'normal' if it is normal at all $\mathbf{p}$.
  We say that good $i$ is 'inferior' if there is $\mathbf{p}$ such that good $i$ is inferior at $\mathbf{p}$.
- We say that $u$ is 'normal' if each good is normal.
  We say that $u$ is 'inferior' if there is some good that is inferior. ◇

*Remarks.* 1. We say 'good $i$ is inferior on $\mathscr{M}$', if there exists $\mathbf{p}$ such that good $i$ is inferior on $\mathscr{M}$ at $\mathbf{p}$.
2. It is impossible that a good is inferior and normal.
3. The statement 'good $i$ is not inferior $\Rightarrow$ good $i$ is normal', does not hold. But it holds in case each utility maxisation problem has a unique maximiser. In this case each good (and $u$) is normal or inferior.
4. For $n = 1$ there does not exist an inferior utility function.[12]

---

[9] Here is a proof by contradiction: so suppose $u$ is Giffen. Then there is some price section $\mathscr{P}$ and budget $m$ such that good 1 is Giffen on $\mathscr{P}$ at $m$. Let $p_a, p_b \in \mathscr{P}$ with $p_a < p_b$. Then $\check{x}_1(p_a) < \check{x}_1(p_b)$.

Because of the upper semi-continuity of $u$ the maximum $v(\mathbf{p}; m)$ of $\dagger_{\mathbf{p};m}$ is well-defined. Because $\check{x}_i(p_b) \subseteq [0, m/p_b] \subseteq [0, m/p_a]$, it follows that, $v(p_b; m) < v(p_a; m)$. Because $\check{x}_1(p_a) < \check{x}_1(p_b)$, it follows that $\check{x}_1(p_a) \subseteq [0, m/p_b]$ and therefore $v(p_a; m) < v(p_b; m)$, a contradiction.

[10] See the contribution [2] in this book for an overview of concrete Giffen utility functions with various theoretical properties.

[11] That is, a subset of $\mathbb{R}_+$ with $\#\mathscr{M} \geq 2$.

[12] Here is a proof by contradiction. Suppose $u$ is inferior. Then there is a budget section $\mathscr{M}$ and price $p_1$ such that good 1 is inferior on $\mathscr{M}$ at $p_1$. Let $m_a, m_b \in \mathscr{M}$ with $m_a < m_b$. Then $\hat{x}_1(m_a) > \hat{x}_1(m_b)$. Because $\hat{x}_1(m_a) \subseteq [0, m_a/p_1] \subseteq [0, m_b/p]$, it follows that $v(p; m_a) < v(p; m_b)$.

5. If a good is inferior on $\mathscr{M}$, then $0 \notin \mathscr{M}$. Even: if good $i$ is inferior on $\mathscr{M}$, then $\inf \mathscr{M} > 0$.[13] This does not exclude, as an example of [4] shows,[14] that a good may be inferior on an unbounded budget section.
6. In [1] another definition is used for the normality of a good: a good $i$ is 'normal' if for all $\mathbf{p}$, the function $\inf \hat{x}_i$ is increasing.

**Proposition 1.** *If good $i$ is Giffen on $\mathscr{P}$ at $(\mathbf{p}_{\hat{\imath}}; m)$ and $p_i \in \mathscr{P}$, then $x_i > 0$ for every maximiser $\mathbf{x}$ of $\dagger_{\mathbf{p};m}$.* ◇

*Proof.* If there exists $p \in \mathscr{P}$ with $p < p_i$, then because good $i$ is Giffen, $\check{x}_i(p) < \check{x}_i(p_i)$ holds which implies that $x_i > 0$. Now suppose there does not exist $p \in \mathscr{P}$ with $p < p_i$.

We will prove by contradiction that $x_i > 0$. So suppose $x_i = 0$. Fix $p \in \mathscr{P}$ with $p > p_i$ and, with $\mathbf{p}' = (p; \mathbf{p}_{\hat{\imath}})$ a maximiser $\mathbf{x}'$ of $\dagger_{\mathbf{p}';m}$. Because $p_i < p$, we have $\mathbf{p} \cdot \mathbf{x}' \leq \mathbf{p}' \cdot \mathbf{x}' \leq m$ and therefore $u(\mathbf{x}') \leq u(\mathbf{x})$. Because $x_i = 0$, we have $\mathbf{p}' \cdot \mathbf{x} = \mathbf{p} \cdot \mathbf{x} \leq m$ and therefore $u(\mathbf{x}) \leq u(\mathbf{x}')$. So we have $u(\mathbf{x}) = u(\mathbf{x}')$. This yields that $\mathbf{x}'$ is also a maximiser of $\dagger_{\mathbf{p};m}$. As good $i$ is also Giffen on $\{p_i, p\}$, it follows that $x_i{}' < x_i{}'$ which is a contradiction. □

## 4 Each Giffen Good Is Inferior

The next proposition provides a strong form of the so-called 'compensated law of demand'. However, the first part of this proposition is the most fundamental and will be used for proving that each Giffen good is inferior.

**Proposition 2.** *Fix a good $i$, budget $m$ and prices $\mathbf{p}$ en $\mathbf{p}'$ with $\mathbf{p}_{\hat{\imath}} = \mathbf{p}'_{\hat{\imath}}$. Let*

$$\mathbf{x} \text{ be a maximiser of } \dagger_{\mathbf{p};m},$$

$$\mathbf{x}' \text{ be a maximiser of } \dagger_{\mathbf{p}';m},$$

*and with* $m' := m + (p_i{}' - p_i)x_i$,[15]

$$\mathbf{x}'' \text{ a maximiser of } \dagger_{\mathbf{p}';m'}.$$

Because $\hat{x}_1(m_b) < \hat{x}_1(m_a)$, it follows that $\hat{x}_1(m_b) \subseteq [0, m_a/p]$ and therefore $v(p; m_b) < v(p; m_a)$, a contradiction.

[13]Here is a proof by contradiction. Suppose $\inf \mathscr{M} \leq 0$. Then $\inf \mathscr{M} = 0$. Take $m_0 \in \mathscr{M}$ with $\hat{x}_i(m_0) > 0$. Because $\inf \mathscr{M} = 0$ and $\min(p_i \hat{x}_i(m_0), m_0) > 0$, there is $m \in \mathscr{M}$ with $m < \min(p_i \hat{x}_i(m_0), m_0)$. Now $p_i \hat{x}_i(m_0) > m \geq p_i \cdot \hat{x}_i(m) > p_i \hat{x}_i(m_0)$, a contradiction.

[14]Also see Example 8 in the contribution [2] in this book.

[15]Note that $m' \geq 0$.

*Then:*

1. $(p_i{}' - p_i)(x_i{}'' - x_i) \geq 0 \;\Rightarrow\; \mathbf{x}''$ *is a maximiser of* $\dagger_{\mathbf{p};m}$.
2. *(Compensated law of demand.) If the maximiser* $\mathbf{x}$ *is unique, then*

$$(p_i{}' - p_i)(x_i{}'' - x_i) \leq 0. \qquad \diamond$$

*Proof.* 1. Note that $\mathbf{p}'\cdot\mathbf{x} \leq m'$. Indeed: $\mathbf{p}'\cdot\mathbf{x} = (\mathbf{p}'-\mathbf{p})\cdot\mathbf{x}+\mathbf{p}\cdot\mathbf{x} = (p_i{}'-p_i)\cdot x_i+\mathbf{p}\cdot\mathbf{x} \leq (p_i{}' - p_i)\cdot x_i + m = m'$. Because $\mathbf{x}''$ is a maximiser of $\dagger_{\mathbf{p}';m'}$, it follows that

$$u(\mathbf{x}) \leq u(\mathbf{x}'').$$

Now with $\mathbf{p}'\cdot\mathbf{x}'' \leq m'$

$$\begin{aligned}\mathbf{p}\cdot\mathbf{x}'' - m &= (\mathbf{p}-\mathbf{p}')\cdot\mathbf{x}'' + \mathbf{p}'\cdot\mathbf{x}'' - m = (p_i - p_i{}')x_i{}'' + \mathbf{p}'\cdot\mathbf{x}'' - m\\ &\leq (p_i - p_i{}')x_i{}'' + m' - m = -(p_i{}' - p_i)(x_i{}'' - x_i) \leq 0.\end{aligned}$$

Therefore

$$\mathbf{p}\cdot\mathbf{x}'' \leq m.$$

Because $\mathbf{x}$ is a maximiser of $\dagger_{\mathbf{p};m}$, the last inequality implies

$$u(\mathbf{x}) \geq u(\mathbf{x}'').$$

So $u(\mathbf{x}) = u(\mathbf{x}'')$ and thus $\mathbf{x}''$ is a maximiser of $\dagger_{\mathbf{p};m}$.

2. If $x_i{}'' = x_i$, then the desired inequality holds. Now suppose $x_i{}'' \neq x_i$. This implies that $\mathbf{x}''$ is not a maximiser of $\dagger_{\mathbf{p};m}$. Now apply 1. □

Because the role of $\mathbf{x}$, $\mathbf{p}$ and $\mathbf{x}'$, $\mathbf{p}'$ in Proposition 2 is the same, we also have:

**Corollary 1.** *Fix a good i, budget m and prices* $\mathbf{p}$ *en* $\mathbf{p}'$ *with* $\mathbf{p}_{\hat{\imath}} = \mathbf{p}'_{\hat{\imath}}$. *Let*

$$\mathbf{x} \textit{ be a maximiser of } \dagger_{\mathbf{p};m},$$

$$\mathbf{x}' \textit{ be a maximiser of } \dagger_{\mathbf{p}';m},$$

*and with* $m' := m + (p_i - p_i{}')x_i{}'$,

$$\mathbf{x}'' \textit{ a maximiser of } \dagger_{\mathbf{p};m'}.$$

*Then:*

1. $(p_i - p_i{}')(x_i{}'' - x_i{}') \geq 0 \;\Rightarrow\; \mathbf{x}''$ *is a maximiser of* $\dagger_{\mathbf{p}';m}$.
2. *(Compensated law of demand.) If the maximiser* $\mathbf{x}'$ *is unique, then*

$$(p_i - p_i{}')(x_i{}'' - x_i{}') \leq 0. \qquad \diamond$$

**Theorem 1.** *Each Giffen good is inferior.*

*More precisely: suppose good $i$ is Giffen on a price section $\mathscr{P}$ at $(\mathbf{p}_{\hat{i}}; m)$. Fix $p_i \in \mathscr{P}$. Then:*

1. *For every $p \in \mathscr{P}$ with $p < p_i$, there exists a budget $m_-(p)$ with $m_-(p) < m$ such that good $i$ is inferior on $\{m_-(p), m\}$ at $(p; \mathbf{p}_{\hat{i}})$. If $\mathbf{x}(p)$ is a maximiser of $\dagger_{(p_i;\mathbf{p}_{\hat{i}});m}$, then the budget*

$$m_-(p) := m + (p - p_i)x_i(p)$$

*is such.*

2. *For every $p \in \mathscr{P}$ with $p < p_i$, there exists a budget $m_+(p)$ with $m_+(p) > m$ such that good $i$ is inferior on $\{m, m_+(p)\}$ at $(p_i; \mathbf{p}_{\hat{i}})$. If $\mathbf{x}'(p)$ is a maximiser of $\dagger_{(p;\mathbf{p}_{\hat{i}});m}$, then the budget*

$$m_+(p) := m + (p_i - p)x_i{}'(p)$$

*is such.* ◇

*Proof.* 1. By Proposition 1,

$$x_i(p) > 0.$$

This implies $m_-(p) < m$. Even

$$0 \leq m_-(p) < m.$$

We now prove by contradiction that good $i$ is inferior on $\{m_-(p), m\}$ at $(p; \mathbf{p}_{\hat{i}})$. So suppose $\hat{x}_i(m_-(p)) > \hat{x}_i(m)$ does not hold. There exists

$$\text{a maximiser } \mathbf{x}'' \text{ of } \dagger_{(p;\mathbf{p}_{\hat{i}});m_-(p)},$$
$$\text{and a maximiser } \mathbf{x}' \text{ of } \dagger_{(p;\mathbf{p}_{\hat{i}});m}$$

such that

$$x_i{}'' \leq x_i{}'.$$

As good $i$ is Giffen on $\{p, p_i\}$ at $(\mathbf{p}_{\hat{i}}; m)$, $\mathbf{x}'$ is a maximiser of $\dagger_{(p;\mathbf{p}_{\hat{i}});m}$ and $\mathbf{x}(p)$ is a maximiser of $\dagger_{(p_i;\mathbf{p}_{\hat{i}});m}$, it follows that $x_i{}' < x_i(p)$ holds. So $x_i{}'' < x_i(p)$ and from this

$$(p - p_i)(x_i{}'' - x_i(p)) > 0.$$

By Proposition 2(1), $\mathbf{x}''$ is a maximiser of $\dagger_{(p_i;\mathbf{p}_{\hat{i}});m}$. Because good $i$ is Giffen on $\{p, p_i\}$ at $(\mathbf{p}_{\hat{i}}; m)$, the inequality

$$x_i{}'' > x_i{}'$$

follows. A contradiction.

2. By Proposition 1,

$$x_i{}'(p) > 0.$$

This yields

$$m_+(p) > m.$$

We now prove by contradiction that good $i$ is inferior on $\{m, m_+(p)\}$ at $(p_i; \mathbf{p}_{\hat{i}})$. So suppose that $\hat{x}_i(m_+(p)) < \hat{x}_i(m)$ does not hold. This implies that there exists

$$\text{a maximiser } \mathbf{x}'' \text{ of } \dagger_{(p_i;\mathbf{p}_{\hat{i}});m_+(p)},$$
$$\text{a maximiser } \mathbf{x} \text{ of } \dagger_{(p_i;\mathbf{p}_{\hat{i}});m}$$

such that

$$x_i \le x_i{}''.$$

As good $i$ is Giffen on $\{p, p_i\}$ at $(\mathbf{p}_{\hat{i}}; m)$, $\mathbf{x}$ is a maximiser of $\dagger_{(p_i;\mathbf{p}_{\hat{i}});m}$ and $\mathbf{x}'(p)$ is a maximiser of $\dagger_{(p;\mathbf{p}_{\hat{i}});m}$, it follows that $x_i > x_i{}'(p)$ holds. So $x_i{}'(p) < x_i{}''$ and from this

$$(p_i - p)(x_i{}'' - x_i{}'(p)) \ge 0.$$

By Corollary 1(1), $\mathbf{x}''$ is a maximiser of $\dagger_{(p;\mathbf{p}_{\hat{i}});m}$. Because good $i$ is Giffen on $\{p, p_i\}$ at $(\mathbf{p}_{\hat{i}}; m)$, the inequality

$$x_i{}'' < x_i$$

follows. A contradiction. □

So Theorem 1 implies in case where each utility maximisation problem has a unique maximiser, that each normal good is ordinary.

If the utility function $u$ is linear homogeneous, then

$$\tilde{\mathbf{x}}(\mathbf{p}; tm) = t\tilde{\mathbf{x}}(\mathbf{p}; m) \ (t \ge 0)$$

holds for all $\mathbf{p}$ and $m$. This implies that no good is inferior. Therefore:

**Corollary 2.** *If a utility function u is a strictly increasing transformation of a linear homogeneous function, then u is neither inferior nor Giffen.* ◇

## 5 Non-decreasing Demand Functions That Are Nowhere Strictly Increasing

The next proposition concerns our construction of a utility function.

**Proposition 3.** *Given positive $a, b, m$ with $a < b$ and a function $f_1 : [a, b] \to \mathbb{R}$ of bounded variation such that*

$$0 < f_1(p_1) < \frac{m}{p_1} \ (a \le p_1 \le b).$$

1. *There exists a function* $u : \mathbb{R}^2_+ \to \mathbb{R}$ *such that for every* $p_1 \in [a,b]$ *the maximisation problem*

$$\mathrm{MAX}_{\substack{(x_1,x_2)\in\mathbb{R}^2_+ \\ p_1x_1+x_2\leq m}} u(x_1,x_2)$$

*has a unique maximiser* $\check{\mathbf{x}}(p_1)$ *with* $\check{x}_1(p_1) = f_1(p_1)$*, and this maximiser is budget-balanced.*
2. *If* $f_1$ *is continuous, then the function u in 1 can be chosen continuous.* ◇

*Proof.* 1. Define $f_2 : [a,b] \to \mathbb{R}$ and $f : [a,b] \to \mathbb{R}^2$ by

$$f_2(p_1) := m - p_1 f_1(p_1), \quad f(p_1) := (f_1(p_1), f_2(p_1)).$$

Note that $0 < f_2 < m$. With

$$H_{p_1} := \{(x_1,x_2) \in \mathbb{R}^2_+ \mid p_1x_1 + x_2 = m\}$$

one has[16] for $p_1 \in [a,b]$

$$f(p_1) \in H_{p_1} = I_{[(0,m),f(p_1)]} \cup I_{[f(p_1),(m/p_1,0)]}.$$

Let

$$G := \cup_{p_1\in[a,b]} H_{p_1}$$

So

$$f([a,b]) \subseteq G \setminus \{(0,m)\}.$$

Note that $p_1 \neq p_1' \Rightarrow H_{p_1} \cap H_{p_1'} = \{(0,m)\}$, so the function $p : G \setminus \{(0,m)\} \to [a,b]$ is well-defined by

$$p(x_1,x_2) := \text{ the unique element } p_1 \in [a,b] \text{ for which } (x_1,x_2) \in H_{p_1}.$$

Also note that

$$p(f(p_1)) = p_1 \ (a \leq p_1 \leq b).$$

Let $L$ be the indefinite variation of $f_1$, i.e. the function $L : [a,b] \to \mathbb{R}$ defined by

$$L(p_1) := \mathrm{Var}_{[a,p_1]} f_1.$$

$L$ is strictly increasing and continuous.

Having all this, we now define $u : \mathbb{R}^2_+ \to \mathbb{R}$. First we define $u \upharpoonright G$:

- $u(0,m) = 1$ and $u = 1$ on $[m/b, m/a] \times \{0\}$.
- For $p_1 \in [a,b]$, define $u$ on $H_{p_1} = I_{[(0,m),f(p_1)]} \cup I_{[f(p_1),(m/p_1,0)]}$ further as follows:

---

[16]We use for $\mathbf{x}, \mathbf{y} \in \mathbb{R}^2$ the notation $I_{[\mathbf{x},\mathbf{y}]} := \{(1-\lambda)\mathbf{x} + \lambda\mathbf{y} \mid 0 \leq \lambda \leq 1\}$.

- $u(f(p_1)) = L(b) - L(p_1) + 2$.
- Noting that $u(0,m) = u(m/p_1, 0) < u(f(p_1))$ $(a \leq p_1 \leq b)$, $u$ is affine and strictly increasing on $I_{[(0,m),f(p_1)]}$ and $u$ is affine and strictly decreasing on $I_{[f(p_1),(m/p_1,0)]}$.

It is clear that we can extend $u \restriction G$ to a function on $\mathbb{R}^2_+$ with $u \restriction \{\mathbf{x} \mid bx_1 + p_2x_2 < m\} < 2$. By construction, for every $p_1 \in [a,b]$ the maximisation problem

$$\text{MAX}_{\substack{\mathbf{x}\in\mathbb{R}^2_+ \\ p_1x_1+x_2=m}} u(\mathbf{x})$$

has $f(p_1)$ as unique maximiser. Now take $\mathbf{x} \in \mathbb{R}^2_+$ with $p_1x_1 + x_2 < m$. If $\mathbf{x} \notin G$, then $bx_1 + x_2 < m$ and therefore $u(\mathbf{x}) < 2 \leq u(f(p_1))$. Now suppose $\mathbf{x} \in G$. Then $\mathbf{x} \in H_{p(\mathbf{x}}$ and $p(\mathbf{x}) > p_1$. It follows that $u(\mathbf{x}) \leq u(f(p(\mathbf{x})) = L(b) - L(p(\mathbf{x})) + 2 < L(b) - L(p_1) + 2 = u(f(p_1))$. Thus also the maximisation problem

$$\text{MAX}_{\substack{\mathbf{x}\in\mathbb{R}^2_+ \\ p_1x_1+x_2\leq m}} u(\mathbf{x})$$

has $f(p_1)$ as unique maximiser and $f(p_1)$ is, by construction, budget-balanced.

2. First we prove here that, with $u$ as in the proof of 1, $u \restriction G$ is continuous. $G = G_- \cup G_+$, where

$$G_+ := \cup_{p_1\in[a,b]} I_{[(0,m),f(p_1)]}, \quad G_- := \cup_{p_1\in[a,b]} I_{[f(p_1),(m/p_1,0)]}.$$

As $G_+$ and $G_-$ are compact, it is sufficient to show that $u \restriction G_+$ and $u \restriction G_-$ are continuous. We now prove that $u \restriction G_-$ is continuous; the proof for $u \restriction G_+$ is analogous.

*Proof.* fix $(x, y) \in G_-$; we prove that $u \restriction G_-$ is continuous in $(x, y)$. Let $\epsilon > 0$. Noting that $x \neq 0$, let $\delta := \frac{x}{1+b}\epsilon$. Let $(x', y') \in G_-$ with $|x - x'|, |y - y'| < \delta$. For $p_1 = p(x, y)$ and $p'_1 = p(x', y')$ we have $|p_1 - p'_1| < \epsilon$. Indeed: $x|p_1 - p'_1| \leq |p_1x - p'_1x'| + |p'_1x' - p'_1x| = |y - y'| + p'_1|x - x'| < (1 + p'_1)\delta \leq (1 + b)\delta = x\epsilon$. Thus

$$\lim_{(x',y')\to(x,y)} p(x', y') = p(x, y).$$

For $(x, y) \in G_-$ we have

$$(x, y) = s\left(\frac{m}{p(x, y)}, 0\right) + (1 - s)f(p(x, y)) \text{ for some } 0 \leq s \leq 1$$

which implies

$$s(x, y) = \frac{x - f_1(p(x, y))}{\frac{m}{p(x,y)} - f_1(p(x, y))}.$$

Because $f_1$ is continuous, the above implies that $s$ continuously depends on $(x, y) \in G_-$. Because

$$u(x,y) = s(x,y)u\left(\frac{m}{p(x,y)}\right) + (1 - s(x,y))u(f_1(p(x,y)))$$
$$= s(x,y) + (1 - s(x,y))u(f_1(p(x,y)))$$

it follows that $u \upharpoonright G_-$ is continuous. It is clear that we can extend $u \upharpoonright G$ to a continuous function on $\mathbb{R}^2_+$ with $u \upharpoonright \{\mathbf{x} \mid bx_1 + p_2x_2 < m\} < 2$. □

In the next lemma we use the fact that there exists a function $F : \mathbb{R} \to \mathbb{R}$ with the following properties[17]:

1. $F$ is differentiable;
2. Both $\{x \in \mathbb{R} \mid F'(x) > 0\}$ and $\{x \in \mathbb{R} \mid F'(x) < 0\}$ are dense in $\mathbb{R}$;
3. $F$ is the difference of two monotone functions.

This deep result is due to [3]; also see Example 13.2 in [6]. Property 2 implies that $F$ is nowhere monotone.

Let $[a,b]$ be a segment in $\mathbb{R}$ with $0 < a < b$ and let $m > 0$. We only consider the part of $F$ defined on $[a,b]$. Without effecting the above three properties, we may assume a number of extra properties for $F$. For example, we may assume that $F(a) < F(b)$.[18] Moreover, by adding a large enough constant, we may assume that $F > 0$, and by multiplying this result by a small enough number that $0 < F(x) < m/x$ $(a \leq x \leq b)$. Thus we have proved the following:

**Lemma 1.** *Given $a,b,m > 0$ with $a < b$, there exists a nowhere monotone differentiable function $f_1 : [a,b] \to \mathbb{R}$ of bounded variation such that*

$$0 < f_1(a) < f_1(b),$$
$$0 < f_1(p_1) < m/p_1 \ (a \leq p_1 \leq b).$$ ◇

**Theorem 2.** *For every $a,b,m > 0$ with $a < b$, there exists a continuous utility function $u : \mathbb{R}^2_+ \to \mathbb{R}$ with the following properties:*

1. *For every $p_1 \in [a,b]$ the utility maximalisation problem $\dagger_{(p_1,1);m}$ has a unique maximiser $\breve{\mathbf{x}}(p_1)$ and this maximiser is budget-balanced.*
2. *The function $\breve{x}_1 : [a,b] \to \mathbb{R}$ is differentiable and of bounded variation.*
3. *Good i is Giffen on $\{a,b\}$ at the price-budget pair $(1;m)$, i.e. $\breve{x}_1(a) < \breve{x}_1(b)$.*
4. *There does not exist $c,d \in \mathbb{R}$ with $c < d$ and $[c,d] \subseteq [a,b]$ such that good i is Giffen on $[c,d]$ at $(1;m)$, i.e. such that $\breve{x}_1 \upharpoonright [c,d]$ is strictly increasing.* ◇

---

[17] Even with the following additional properties: $|F'(x)| \leq 1$ $(x \in \mathbb{R})$, $\{x \in \mathbb{R} \mid F'(x) = 0\}$ is dense in $\mathbb{R}$ and $F'$ is over no interval Riemann-integrable.

[18] For if $F(a) > F(b)$, then replace $F$ by $-F$. And if $F(a) = F(b)$, there is always a constant $c$ with $a < c < b$ with $F(a) \neq F(c)$. Now replace $F(x)$ by $\tilde{F}(x) := \left(a\frac{b-c}{b-a} + \frac{c-a}{b-a}x\right)$. Now $\tilde{F}(a) = F(a) \neq F(c) = \tilde{F}b$. If $\tilde{F}(a) > \tilde{F}(b)$, then replace $\tilde{F}$ by $-\tilde{F}$.

*Proof.* Fix $a, b, m > 0$ with $a < b$. Let $f_1$ be as in Lemma 1. By Proposition 3(2), there exists a continuous function $u : \mathbb{R}^2_+ \to \mathbb{R}$ such that for every $p_1 \in [a, b]$ the maximisation problem $\dagger_{(p_1,1);m}$ has a unique maximiser $\breve{\mathbf{x}}(p_1)$ with $\breve{x}_1(p_1) = f_1(p_1)$. So properties 1–3 hold. Because $f_1 \upharpoonright [c, d]$ is not increasing, it also is not strictly increasing and thus property 4 holds. □

*Remark.* In Theorem 2 we only consider prices $p_1 \in [a, b]$. The form of $\breve{x}_1(p_1)$ outside $[a, b]$ depends on how one defines $u$ outside $G$. It is clear that one can do this in such a way that one obtains a well-defined demand function $\breve{x}_1 : \mathbb{R}_{++} \to \mathbb{R}$ such that $\breve{x}_1 \upharpoonright ]0, a]$ and $\breve{x}_1 \upharpoonright [b, \infty[$ are strictly decreasing. Theorem 2(4) then implies that there does not exist a non-degenerate interval $I$ of $\mathbb{R}$ with $I \subseteq \mathbb{R}_{++}$ such that $\breve{x}_1 \upharpoonright I$ is strictly increasing. So we have here a situation where (1) holds, but (2) does not hold.

The next lemma shows that Theorem 2 does not hold anymore if we replace its second property by '$\breve{x}_1 : [a, b] \to \mathbb{R}$ is continuously differentiable'.

**Lemma 2.** *Let $a < b$ and $f : [a, b] \to \mathbb{R}$ a continuously differentiable function with $0 < f(a) < f(b)$. Then there exists a non-degenerate interval $I$ of $\mathbb{R}$ with $I \subseteq ]a, b[$ such that $f(x) \in ]f(a), f(b)[ (x \in I)$ and $f \upharpoonright I$ strictly increasing (even $f'(x) > 0$ $(x \in I)$).* ◇

*Proof.* It will be proved that there exists $c \in ]a, b[$ with $f(c) \in ]f(a), f(b)[$ and $f'(c) > 0$. This then implies the desired results. Indeed: because $f'$ is continuous in $c$, there exists $\xi_1 > 0$ with $]c - \xi_1, c + \xi_1[ \subseteq ]a, b[$ such that $f' > 0$ on $]c - \xi_1, c + \xi_1[$ Because $f$ is continuous in $c$, there exists $\xi_2 \in ]0, \xi_1[$ such that $f > 0$ on $]c - \xi_2,\ c + \xi_2[$. Now $I = ]c - \xi_2, c + \xi_2[$ is as desired.

Consider two cases.

- Case $f'(b) > 0$. Because $f'$ is continuous in $b$, there exists $\delta_1 > 0$ with $b - \delta_1 > a$ such that $f' > 0$ on $[b - \delta_1, b]$. Thus $f$ is strictly increasing on $[b - \delta_1,\ b]$. Because $f(b) > f(a)$ and $f$ is continuous in $b$, this implies that there exists $\delta_2 \in ]0, \delta_1]$ such that $f(x) \in ]f(a), f(b)[$ on $[b - \delta_2, b[$. Thus every $c \in [b - \delta_2, b[$ is as desired.
- Case $f'(b) \leq 0$. Let

$$m := \frac{f(b) - f(a)}{b - a}.$$

Note $m > 0$. Define $g : [a, b] \to \mathbb{R}$ by

$$g(x) := f(x) - (mx + f(a) - ma).$$

One has $g(a) = g(b) = 0$. Also $g'(b) = f'(b) - m < 0$. Due to the continuity of $g'$, there exists $\delta > 0$ such that $g' < 0$ on $W = [b - \delta, b]$ and thus $g$ is strictly decreasing on $W$. It follows that $g > 0$ on $W$.

The set $\{x \in W \mid g(x) = 0\}$ is, because $W$ is compact and $g$ is continuous, compact. Because $a \in W$, it is non-empty and therefore its maximum $c$ is well-defined. Of course, $c \neq b$, $g(c) = 0$ and $f(c) = g(c) + (mc + f(a) - ma) = f(a) + m(c - a) \in [f(a), f(b)[$.

Because $g(c) = g(b) = 0$ and $g$ is continuous, it follows that $g > 0$ on $]c, b[$. Because $g$ is continuously differentiable, it follows that $g'(c) \geq 0$ and thus that $f'(c) \geq m > 0$.

If $c \neq a$, the proof is complete. If $c = a$, then, because $f'(c) > 0$ and $f$ is continuous differentiable, there exists $\eta > 0$ with $a + \eta < b$, such that $f'(a + \eta) > 0$ and $f(a + \eta) \in ]f(a), f(b)[$. Now $a + \eta$ is as desired. □

## References

1. I. Dasgupta and P. K. Pattanaik. Comparative statics for a consumer with multiple optimum consumption bundles. IZA Discussion Paper 4818, Institute for the Study of Labor, Bonn, 2010.
2. W. Heijman and P. H. M. von Mouche. A child garden of concrete Giffen utility functions: a theoretical review. 655:69–88, 2011.
3. Y. Katznelson and K. Stromberg. Everywhere differentiable, nowhere monotone, functions. *American Mathematical Monthly*, 81:349–353, 1974.
4. H. Liebhafsky. New thoughts about inferior goods. *American Economic Review*, 59:931–934, 1969.
5. A. Mas-Colell, M. D. Whinston, and J. R. Green. *Microeconomic Theory*. Oxford University Press, New York, 1995.
6. A. van Rooy and W. Schikhof. *A Second Course on Real Functions*. Cambridge University Press, Cambridge, 1982.
7. P. Sørenson. Giffen demand for several goods. 655:97–104, 2011.

# Giffen Behaviour and Strong Asymmetric Gross Substitutability

**Kris De Jaegher**

**Abstract** Let a consumer consume two goods, and let good 1 be a Giffen good. Then a well-known necessary condition for such behaviour is that good 1 is an inferior good. This paper shows that an additional necessary condition for such behaviour is that good 1 is a gross substitute for good 2, and that good 2 is a gross complement to good 1 (strong asymmetric gross substitutability). It is argued that identifying strong asymmetric gross substitutability as an additional necessary condition gives better insight into Giffen behaviour, both on an analytical level and an intuitive level.

## 1 Introduction

Any paper that treats Giffen behaviour may be confronted with the criticism: to what extent is such behaviour relevant? While there is empirical evidence for Giffen behaviour ([1] and [7]), one may still argue that the law of demand is violated only infrequently. Yet, the point of Giffen behaviour is that it is a tool to get insight into what determines the price elasticity of demand. The fact that the income effect may exceed the substitution effect helps remind us that these two determining effects may not operate in the same direction. Also, even if one accepts Giffen goods do not occur very frequently, there still remains a class of goods that are very similar to Giffen goods, in that their price elasticity of demand is negative but very small, or is even zero. Getting insight into the extreme possibility of Giffen goods also gives us insight into these similar goods, which in this reasoning are important as they occur more frequently.

K. De Jaegher (✉)
Utrecht University School of Economics, P.O. Box 80125, 3508TC Utrecht, Netherlands
e-mail: k.dejaegher@uu.nl

W. Heijman and P. Mouche (eds.), *New Insights into the Theory of Giffen Goods*, Lecture Notes in Economics and Mathematical Systems 655,
DOI 10.1007/978-3-642-21777-7_5, 

In Sect. 2, we derive a list of necessary (though not sufficient) conditions for Giffen behaviour. This includes well-known necessary conditions such as non-homothetic preferences, and the fact that the Giffen good must also be an inferior good while the other good consumed is a luxury (strong asymmetric income elasticity), which we refer to as strong asymmetric income elasticity. To these well-known conditions, we add the necessary conditions of weak asymmetric gross substitutability, where the cross-price effects on the two goods are unequal, and of strong asymmetric gross substitutability [4], where good 1 is a gross substitute for good 2, while good 2 is a gross complement to good 1.

In Sect. 3, we provide an example suggesting that both strong asymmetric income elasticity and strong asymmetric gross substitutability are needed for it to be possible that there is Giffen behaviour. Put otherwise, if preferences are only characterised by one of these two strong asymmetries, there cannot be Giffen behaviour. This shows that strong asymmetric gross substitutability is as essential for Giffen behaviour as is strong asymmetric income elasticity. We end with a discussion in Sect. 4.

## 2 Derivation of Necessary Conditions

In this section, we treat several necessary conditions for Giffen behaviour. We focus on the case where two goods are consumed, and where good 1 is the potential Giffen good. The analysis is easily extended to multiple good when good 2 is thought of as a composite good.[1] In the context of this book, it is important to not that we will not always make all underlying assumptions explicit. For example, when we apply the Slutsky equation, we simply assume that the conditions necessary for the application of this equation are valid. Throughout, we assume budget-balancedness of maximisers.

The necessary conditions we derive all boil down to some form of asymmetry in the effect of income and price changes on the consumption of the two goods. The necessary conditions are ordered from weak necessary conditions to stronger ones. Section 2.1 shows that good 1 should be a necessity, and good 2 a luxury. Section 2.2 shows that good 1 should be a relatively better substitute for good 2 than good 2 is for good 1. Section 2.3 repeats the well-known necessary condition that good 1 should not only be a necessity but should even be an inferior good. Section 2.4 shows that good 1 should be a gross substitute for good 2, while good 2 should be a gross complement to good 1 (strong asymmetric gross substitutability). Section 2.5 finally shows that good 1 should be price inelastic, while good 2 should be price elastic (strong asymmetric price elasticity).

[1] Specifically, by Hicks' Composite Commodity Theorem (for a proof, see Carter [2]), all other goods than good 1 can be treated as a composite good as long as the relative prices of all these other goods do not change. A change in the price of good 2 is then interpreted as a price change that leaves the relative prices of all other goods than good 1 intact.

## 2.1 Unequal Income Elasticities: Weak Asymmetric Income Elasticity

A necessity (luxury) is defined as good $i$ for which the income elasticity, denoted $\varepsilon_{x_i m}$, is smaller (larger) than one: $\varepsilon_{x_i m} < 1$ ($\varepsilon_{x_i m} > 1$). Define as weak asymmetric income elasticity the case where one good is a necessity, and the other good is a luxury.

**Definition 1.** Two goods 1 and 2 are locally characterised by *weak asymmetric income elasticity* if it is locally the case that $\varepsilon_{x_1 m} < 1$ and $\varepsilon_{x_2 m} > 1$.

This leads us to the following proposition.

**Proposition 1.** *A necessary condition for good 1 to be locally a Giffen good (i.e. locally $\partial x_1/\partial p_1 > 0$) is that preferences are locally characterised by weak asymmetric income elasticity, where locally good 1 is a necessity and good 2 a luxury.*

*Proof.* As is well-known, the Hicksian decomposition of the own price-derivatives in a substitution effect and an income effect takes the form

$$\frac{\partial x_1}{\partial p_1} = \frac{\partial h_1}{\partial p_1} - x_1 \frac{\partial x_1}{\partial m} \tag{1}$$

$$\frac{\partial x_2}{\partial p_2} = \frac{\partial h_2}{\partial p_2} - x_2 \frac{\partial x_2}{\partial m} \tag{2}$$

where $h_i$ denotes the compensated Hicksian demand for good $i$. As $\partial h_1/\partial p_1 \leq 0$, it follows that a necessary condition for good 1 to be a Giffen good is that $\partial x_1/\partial m < 0$, meaning that a necessary condition is that $\varepsilon_{x_1 m} < 1$, or good 1 is a necessity.

Furthermore, by the Engel aggregation identity, we have

$$s_1 \varepsilon_{x_1 m} + s_2 \varepsilon_{x_2 m} = 1 \tag{3}$$

where $s_i = (p_i x_i)/m$ is the share of the budget that the consumer spends on good $i$. By budget-balancedness, $s_1 + s_2 = 1$. It follows from this that if good 1 is a necessity ($\varepsilon_{x_1 m} < 1$), good 2 must be a luxury ($\varepsilon_{x_2 m} > 1$).

Another way to state Proposition 1 is that a necessary condition for Giffen behaviour is that the consumer has *non-homothetic preferences*. Concretely, homothetic utility functions have what we can refer to as *strong symmetric income elasticity* ($\varepsilon_{x_1 m} = \varepsilon_{x_2 m} = 1$), a property which by Proposition 1 should not apply under Giffen behaviour.

**Corollary 1.** *A necessary condition for good 1 to be locally a Giffen good is that the consumer locally has non-homothetic preferences.*

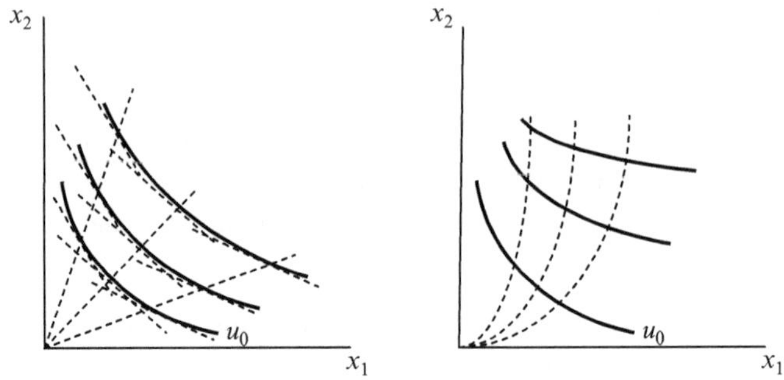


**Fig. 1** Homothetic and non-homothetic preferences

As shown in the left part of Fig. 1, for any given indifference curve, we can find a corresponding homothetic indifference map by "blowing up" the given indifference curve $u_0$. We then have linear income expansion paths, denoted by the dashed upward sloping lines. Put otherwise, for all consumption bundles where the goods are consumed in a fixed proportion to one another (i.e., lying on line through the origin), the marginal rate of substitution is the same, as can be seen from the dashed downward-sloping lines in the left part of Fig. 1. Non-homothetic preferences, where good 1 is a necessity (and possibly an inferior good) and good 2 is a luxury are now obtained by making the blown up versions of the given indifference curve flatter, as illustrated in the right part of Fig. 1. We then obtain non-linear income expansion paths. Graphically, a necessary condition for Giffen behaviour is that indifference curves are not blown-up versions of one another, as illustrated in the left part of Fig. 1. In this figure, good 1 is a necessity, but not an inferior good, as the income expansion paths do not have a negative slope.

## 2.2 Inferior Good/Luxury: Strong Asymmetric Income Elasticity

By the analysis in Sect. 2.1, we can be more precise about the necessary condition referring to income elasticity. Define good $i$ as an inferior good when $\varepsilon_{x_i m} < 0$. We can then additionally introduce the concept of strong asymmetric income elasticity.

**Definition 2.** Two goods 1 and 2 are locally characterised by *strong asymmetric income elasticity* if locally one good is an inferior good ($\varepsilon_{x_1 m} < 0$) and the other good a luxury ($\varepsilon_{x_2 m} > 1$).

Note that strong asymmetric income elasticity (Definition 2) is thus a special case of weak asymmetric income elasticity. A well-known result now allows us to formulate the necessary conditions for Giffen behaviour sharper:

**Proposition 2.** *A necessary condition for good 1 to be locally a Giffen good is that preferences are locally characterised by strong asymmetric income elasticity, where locally good 1 is an inferior good and good 2 a luxury.*

*Proof.* See the proof of Proposition 1.

## 2.3 *Unequal Gross Cross-Price Effects: Weak Asymmetric Gross Substitutability*

Section 2.1 shows that a necessary condition for Giffen behaviour is weak asymmetric income elasticity, meaning that we have non-homothetic preferences. We here show that this asymmetry can be expressed in another form, in the sense that a necessary condition for Giffen behaviour is that goods are *asymmetrically gross substitutable*. The expression $\partial x_i/\partial p_j$ (with $i \neq j$) may be seen as measuring the extent to which good $i$ is a gross substitute for ($\partial x_i/\partial p_j > 0$) or gross complement to ($\partial x_i/\partial p_j < 0$) good $j$. (A)symmetric gross substitutability compares this measure across goods. We start by showing that with homothetic preferences, goods are what we call strongly symmetrically gross substitutable, we first define:

**Definition 3.** Two goods 1 and 2 are locally characterised by *strong symmetric gross substitutability* if $\partial x_1/\partial p_2 = \partial x_2/\partial p_1$.

This leads us to the following proposition:

**Proposition 3.** *The consumer has homothetic preferences over goods 1 and 2 ($\varepsilon_{x_1m} = \varepsilon_{x_2m} = 1$) if and only if it is globally true that $(\partial x_2/\partial p_1) = (\partial x_1/\partial p_2)$.*

*Proof.* Consider the Hicksian decomposition of the cross-price effects:

$$\frac{\partial x_1}{\partial p_2} = \frac{\partial h_1}{\partial p_2} - x_2\frac{\partial x_1}{\partial m} \tag{4}$$

$$\frac{\partial x_2}{\partial p_1} = \frac{\partial h_2}{\partial p_1} - x_1\frac{\partial x_2}{\partial m} \tag{5}$$

As is well-known, given the symmetry of the Slutsky matrix, the Hicksian cross-price effects are equal, or $\partial h_1/\partial p_2 = \partial h_2/\partial p_1$. Combining (4) and (5), we see that $\partial x_2/\partial p_1 = \partial x_1/\partial p_2 \Leftrightarrow \varepsilon_{x_1m} = \varepsilon_{x_2m}$.

The fact that homothetic preferences are equivalent to globally strong symmetric gross substitutability is illustrated in Fig. 2. In Fig. 2a–c, homothetic preferences are each time obtained by constructing blown-up versions of the indifference curve in bold. For a bundle 1 on the bold indifference curve, we construct a line crossing the origin and bundle 1. On every bundle 2 on this line, consumption of each good has increased by the same factor. For an income increase (parallel shift in the budget line through 1) that makes bundle 2 affordable, bundle 2 should be chosen, so

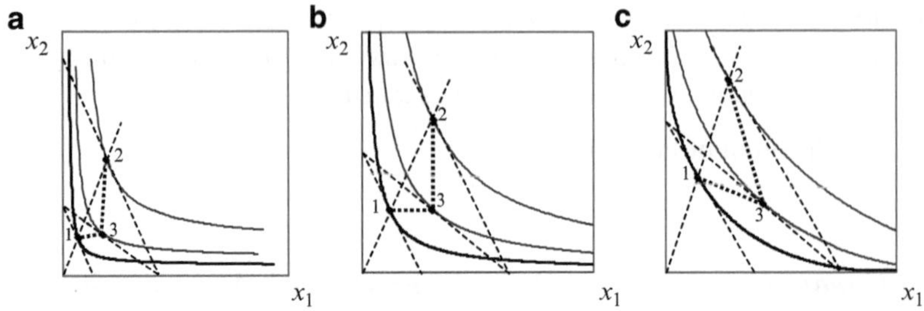


**Fig. 2** Homothetic preferences and strong symmetric gross substitutability

that the indifference curve through bundle 2 must have the same slope. In order to highlight the relation between homotheticity and symmetric substitutability, the income increase from bundle 1 to bundle 2 can be split up in a price decrease in good 1 (from bundle 1 to bundle 3), and an equiproportional price decrease in good 2 (from bundle 3 to bundle 2). This is because two equiproportional price decreases are equivalent to an income increase. In Fig. 2a, each good is a gross complement to the other good, in Fig. 2b, both goods are neither gross complements nor gross substitutes. In Fig. 2c finally, each good is a gross substitute for the other good. The relation between homotheticity and symmetric substitutability is especially clear in Fig. 2b. By the fact that the line connecting bundles 1 and 3 is horizontal, good 2 is neither a substitute for nor a complement to good 1. If the line connecting bundles 3 and 2 would not be vertical, so that it is not the case that good 1 is neither a substitute for nor a complement to good 1, bundle 2 would not lie on the line connecting bundle 1 to the origin, and we would not have homothetic preferences.

We now show that another way to state the necessary condition for Giffen behaviour that there is a necessity and a luxury is to say that the gross cross-price effects on each good are not equal, i.e. that they are not locally characterised by strong symmetric gross substitutability. We refer to this property as *weak asymmetric gross substitutability*.

**Definition 4.** Two goods 1 and 2 are locally characterised by *weak asymmetric gross substitutability* if $\partial x_1/\partial p_2 \neq \partial x_2/\partial p_1$.

This leads us to the following proposition:

**Proposition 4.** *Good 1 is locally a necessity and good 2 is a luxury if and only if locally $\partial x_1/\partial p_2 \neq \partial x_2/\partial p_1$ (= weak asymmetric substitutability). More precisely, good 1 is locally a necessity and good 2 is a luxury if and only if locally $\partial x_2/\partial p_1 < \partial x_1/\partial p_2$.*

*Proof.* Given again that $\partial h_1/\partial p_2 = \partial h_2/\partial p_1$, and combining (4) and (5), we see that $\partial x_2/\partial p_1 < \partial x_1/\partial p_2 \Leftrightarrow \varepsilon_{x_1 m} < \varepsilon_{x_2 m}$. By (3), $\varepsilon_{x_1 m} < \varepsilon_{x_2 m}$ implies that $\varepsilon_{x_1 m} < 1 < \varepsilon_{x_2 m}$.

Given that weak asymmetric income elasticity is a necessary condition for Giffen behaviour and is equivalent to weak asymmetric gross substitutability, the latter is also a necessary condition for Giffen behaviour. In general, a necessary condition for Giffen behaviour is asymmetry in the consumers' preferences. This asymmetry can be stated as weak asymmetric income elasticity, where good 1 is a necessity and good 2 is a luxury. As shown by Proposition 4, this turns out to be the same as saying that the gross cross-price effect of the price of good 2 on the consumption of good 1 is higher than the cross-price effect of the price of good 1 on the consumption of good 2. How to interpret this result intuitively? The case is made here for interpreting this as good 1 being a better gross substitute for good 2 than good 2 is for good 1. Intuitively, let good 1 be bread, and let good 2 be yachts. From the perspective of subsistence, bread is a better substitute for yachts than yachts are for bread (weak asymmetric gross substitutability). You can eat bread, but you cannot eat yachts. There is a one-to-one relationship of this fact with weak asymmetric income elasticity. Given that bread is a necessity, if the price of yachts increases, the income effect of this on bread is relatively small, so that the consumption of bread changes relatively much, as the impact of the substitution effect is only slightly limited by the income effect. Given that yachts are a luxury, if the price of bread increases, the income effect of this on the consumption of yachts is relatively large, so that the consumption of yachts decreases relatively little, as the impact of the substitution effect is considerably limited by the income effect.

The one-to-one relation between weak asymmetric income elasticity and weak asymmetric gross substitutability is shown graphically in Fig. 3. The effect of an income increase from bundle 1 to bundle 2 can be split up in a decrease in the price of good 2 from bundle 1 to bundle 3, plus an equiproportional decrease in the price of good 1 from bundle 3 to bundle 2 or bundle 2'. Bundle 2' is the bundle that is obtained if the consumer has homothetic preferences, and has income expansion paths that are straight lines through the origin. As shown in Proposition 1 and illustrated in Fig. 2, the goods are then equally good gross substitutes. In Fig. 3a, we make preferences non-homothetic by making good 2 a worse gross substitute for good 1, so that higher-up indifference curves become flatter; the income expansion

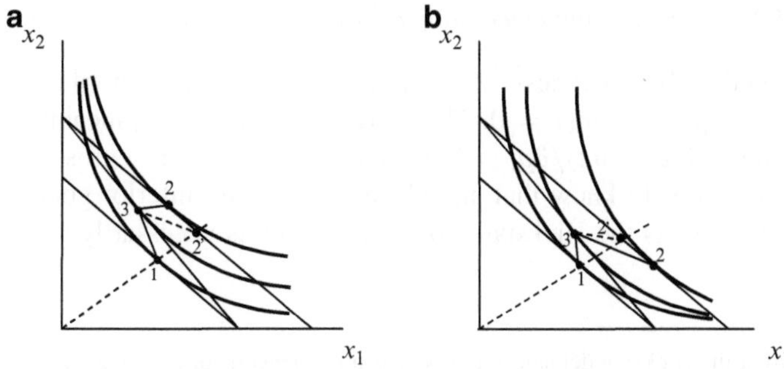


**Fig. 3** Non-homothetic preferences and weak asymmetric gross substitutability

path running through bundles 1 and 2 has a convex shape, so that good 1 is a necessity and good 2 a luxury. We thus move in the direction of good 1 being a Giffen good. In Fig. 3b, we make preferences non-homothetic by making good 2 a better gross substitute for good 1, so that higher-up indifference curves become steeper; the income expansion path running through bundles 1 and 2 is concave so that good 1 is a luxury and good 2 a necessity. We thus move further away from good 1 being a Giffen good.

## 2.4 Good 1 Gross Substitute for Good 2, Good 2 Gross Complement to Good 1: Strong Asymmetric Gross Substitutability

We now show that a more detailed necessary condition for Giffen behaviour is that the gross cross-price effects on each good have opposite signs. Thus, in order to have Giffen behaviour, not only must preferences be asymmetric in the sense that one good 1 is a better gross substitute for good 2 than good 2 is for good 1, they must be asymmetric in the sense that good 1 is a gross substitute for good 2 while good 2 is a gross complement to good 1. We define such a phenomenon as *strong asymmetric substitutability*.

**Definition 5.** Two goods 1 and 2 are locally characterised by *strong asymmetric gross substitutability* if $sign(\partial x_1/\partial p_2) \neq sign(\partial x_2/\partial p_1)$. Two goods 1 and 2 are locally characterised by *weak symmetric gross substitutability* if $sign(\partial x_1/\partial p_2) = sign(\partial x_2/\partial p_1)$.

Note that strong asymmetric gross substitutability (Definition 5) is a special case of weak asymmetric gross substitutability (Definition 4). The concept of strong asymmetric gross substitutability leads us to the following proposition:

**Proposition 5.** *A necessary condition for good 1 to be locally a Giffen good is that the preferences are locally characterised by strong asymmetric gross substitutability, where that good 1 is a gross substitute for good 2 ($\partial x_1/\partial p_2 > 0$), while good 2 is a gross complement to good 1 ($\partial x_2/\partial p_1 < 0$).*

*Proof.* By budget-balancedness, $p_1 x_1 + p_2 x_2 = m$. This implies that $x_1 + p_1(\partial x_1/\partial p_1) + p_2(\partial x_2/\partial p_1) = 0$. Therefore, if $\partial x_1/\partial p_1 > 0$, meaning that good 1 is a Giffen good, then $\partial x_2/\partial p_1 < 0$, meaning that good 2 is a gross complement to good 1. We already know that any Giffen good is an inferior good, so that in (4), $\partial x_1/\partial m < 0$. As the Hicksian cross-price effect is necessarily non-negative[2],

[2]By the fact that the Hicksian demand for good 1 is homogeneous of degree zero, it follows from Euler's formula that $p_1(\partial h_1/\partial p_1) + p_2(\partial h_1/\partial p_2) = 0$. We know that $\partial h_1/\partial p_1 \leq 0$, so that it follows that $\partial h_1/\partial p_2 \geq 0$.

it follows from equation (4) that when good 1 is a Giffen good, good 1 must be a gross substitute for good 2.

Strong asymmetric gross substitutability is not a familiar concept in the literature, and many economists seem to be unaware of its possibility, yet it is perfectly in line with rational preferences. Intuitively, let a family own a smaller car to drive locally in the city, and a larger car for larger trips. The larger car is a complement to the small car; it is not a substitute for the smaller car, because it is too big for parking in the city (imagine that the larger car is one of the bigger four-wheel drives on the market). But the smaller car may at the same time be a substitute for the larger car: if larger cars become too expensive, the family will only keep a smaller car for driving both small and larger distances (imagine that the smaller car is still a reasonable family car, so that larger distances are not that uncomfortable with the smaller car). For other potential examples, see [4].

## 2.5 *Strong Asymmetric Price Elasticity*

As we show in this section, a necessary condition for local Giffen behaviour is that locally, it is the case for good 1 that $\varepsilon_{x_1 p_1} > -1$, and that $\varepsilon_{x_2 p_2} < -1$. We do this by showing that this necessary condition is equivalent to the necessary condition of strong asymmetric substitutability treated in Sect. 2.4. We say that two goods are locally characterised by *weak symmetric elasticity* if both price elasticities are smaller than minus one, or both are larger than minus one.

**Definition 6.** Two goods 1 and 2 are locally characterised by *weak symmetric price elasticity* if locally either $\varepsilon_{x_1 p_1} < -1$ and $\varepsilon_{x_2 p_2} < -1$, or $\varepsilon_{x_1 p_1} > -1$ and $\varepsilon_{x_2 p_2} > -1$.

One could also define the concepts of strong symmetric price elasticity, where both goods have exactly the same elasticity, and weak asymmetric price elasticity, where one good is more elastic than the other, but it need not be the case that one good is elastic and the other inelastic. These concepts, however, tell little about preferences, because relative elasticity depends also on the relative budget shares of each good. This is why we only additionally define the concept of *strong asymmetric price elasticity*.

**Definition 7.** Two goods 1 and 2 are locally characterised by *strong asymmetric price elasticity* if locally either $\varepsilon_{x_1 p_1} < -1$ and $\varepsilon_{x_2 p_2} > -1$, or $\varepsilon_{x_1 p_1} > -1$ (good 1 has a negatively-sloped demand and is price inelastic, or it has a positively-sloped demand) and $\varepsilon_{x_2 p_2} < -1$ (good 2 has a negatively-sloped demand and is price elastic).

We can now state the following proposition, showing equivalence between strong asymmetric price elasticity and strong asymmetric gross substitutability, and between weak symmetric price elasticity and weak symmetric gross substitutability.

**Proposition 6.** *(i) Strong asymmetric price elasticity is equivalent to strong asymmetric gross substitutability.*
*(ii) Weak symmetric price elasticity is equivalent to weak symmetric gross substitutability.*

*Proof.* By budget-balancedness, the consumer always uses up the complete budget, so that it must be the case that $x_i = m/p_i - (p_j/p_i)x_j$ for $i, j = 1, 2$ and $i \neq j$. Taking the derivative of this expression with respect to $p_i$, and expressing in terms of elasticities, we obtain that

$$\varepsilon_{x_1 p_1} = -1 - (s_2/s_1)\varepsilon_{x_2 p_1} \quad (6)$$

$$\varepsilon_{x_2 p_2} = -1 - (s_1/s_2)\varepsilon_{x_1 p_2} \quad (7)$$

By (6) and (7), good 1 is a gross substitute for good 2 ($\varepsilon_{x_1 p_2} > 0$) and good 2 is a gross complement to good 1 ($\varepsilon_{x_2 p_1} < 0$) (i.e., strong asymmetric gross substitutability) if and only if $\varepsilon_{x_1 p_1} > -1$, and $\varepsilon_{x_2 p_2} < -1$. Further, by (6) and (7), both goods are gross complements to one another, meaning that, $\varepsilon_{x_1 p_2} < 0$ and $\varepsilon_{x_2 p_1} < 0$ (weak symmetric gross substitutability), if and only if $\varepsilon_{x_1 p_1} > -1$ and $\varepsilon_{x_2 p_2} > -1$. Given that strong gross asymmetric substitutability is a necessary condition for Giffen behaviour, it follows that both goods are gross complements to one another if and only if $0 > \varepsilon_{x_1 p_1} > -1$ and $0 > \varepsilon_{x_2 p_2} > -1$. Similarly, both goods are gross substitutes to one another, meaning that $\varepsilon_{x_1 p_2} > 0$ and $\varepsilon_{x_2 p_1} > 0$ (weak symmetric gross substitutability), if and only if $\varepsilon_{x_1 p_1} < -1$ and $\varepsilon_{x_2 p_2} < -1$. We have thus established that weak symmetric gross substitutability is equivalent to weak symmetric price elasticity.

**Corollary 2.** *Strong asymmetric price elasticity is a necessary condition for Giffen behaviour. If there is weak symmetric price elasticity, there cannot be Giffen behaviour.*

*Proof.* As by Proposition 5 strong asymmetric gross substitutability is a necessary condition for Giffen behaviour, and as by Proposition 6 strong asymmetric gross substitutability is equivalent to strong asymmetric price elasticity, it follows that strong asymmetric price elasticity is a necessary condition for Giffen behaviour. As by Proposition 5 strong asymmetric gross substitutability is a necessary condition for Giffen behaviour, it follows that if there is weak symmetric gross substitutability and by Proposition 6, with it weak symmetric price elasticity, then there cannot be Giffen behaviour.

A further corollary is now that we can narrow down the case of weak symmetric price elasticity with $\varepsilon_{x_1 p_1} > -1$ and $\varepsilon_{x_2 p_2} > -1$ to the case where the demand for both goods has a negative slope and is price inelastic.

**Corollary 3.** *Let preferences be locally characterised by weak symmetric price elasticity with $\varepsilon_{x_1 p_1} > -1$ and $\varepsilon_{x_2 p_2} > -1$. Then $0 > \varepsilon_{x_1 p_1} > -1$ and $0 > \varepsilon_{x_2 p_2} > -1$.*

**Table 1** Necessary conditions for good 1 to be a Giffen good

| | Good 1 | Good 2 |
|---|---|---|
| Weak asymmetric income elasticity | $\varepsilon_{x_1m} < 1$ | $\varepsilon_{x_2m} > 1$ |
| Weak asymmetric gross substitutability | $\partial x_1/\partial p_2 >$ | $\partial x_2/\partial p_1$ |
| Strong asymmetric income elasticity | $\varepsilon_{x_1m} < 0$ | $\varepsilon_{x_2m} > 1$ |
| Strong asymmetric gross substitutability | $\partial x_1/\partial p_2 > 0$ | $\partial x_2/\partial p_1 < 0$ |
| Strong asymmetric price elasticity | $\varepsilon_{x_1p_1} > -1$ | $\varepsilon_{x_2p_2} < -1$ |

We conclude that necessary conditions for Giffen behaviour are found in several ways in which preferences are asymmetric. One good should be a necessity, and one good should be a luxury, and goods should not be equally-good gross substitutes for each other. The asymmetry should even be more extreme, in that good 1 should be inferior, and good 2 should be a luxury. Also, good 1 should be a gross substitute for good 2, and good 2 a gross complement to good 1, which is exactly equivalent to saying that good 1 should be price inelastic, and good 2 price elastic. The necessary conditions are summarised in Table 1. Note that $\varepsilon_{x_1p_1} > -1$, $\varepsilon_{x_2p_2} < -1$ is used as a necessary condition rather than the necessary (and sufficient) condition $\varepsilon_{x_1p_1} > 0$, $\varepsilon_{x_2p_2} < -1$ because we are interested in wider classes of local preferences that have some of the characteristics of Giffen behaviour. In the next section, we look at the relation between these different necessary conditions.

## 3 Relation Between Necessary Conditions

The next example shows the interrelationship between the different necessary conditions derived in Sect. 2. The argument here is that the sets of preferences locally characterised by strong asymmetric gross substitutability and strong asymmetric income elasticity are not subsets of one another. This suggests that both the necessary condition of strong asymmetric income elasticity *and* the necessary condition of strong asymmetric gross substitutability is essential for Giffen behaviour (though validity of both these necessary conditions is not sufficient). It is possible that one condition is valid, and the other is not, in which case Giffen behaviour is impossible. The relation between the necessary conditions is summarised in Fig. 4. The set of non-homothetic preferences (i.e., the preferences locally characterised by weak asymmetric gross substitutability and weak asymmetric income elasticity) contain two subsets, namely the set of preferences locally characterised by strong asymmetric gross substitutability, and the set of preferences locally characterised by strong asymmetric income elasticity. The latter two subsets intersect. The set of preferences locally characterised by Giffen behaviour is a subset of this intersection.[3] Example 1 does not show that preferences locally characterised by

[3]There are also utility functions that lie on the edge of each of these sets, e.g. preferences that are neither characterised by strong asymmetric gross substitutability, nor by weak symmetric gross

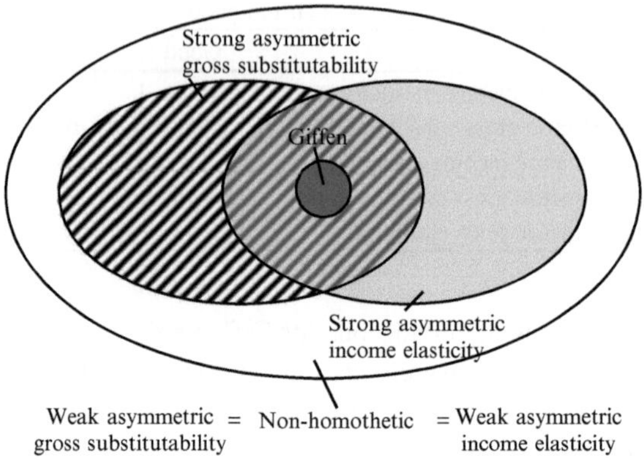


**Fig. 4** Relation between the sets of local preferences meeting the necessary conditions

Giffen behaviour exist – for this the reader may refer to [6] in this book. Example 1, however, suggests that preferences do exist for all the other cases in Fig. 4.

*Example 1.* Consider the continuous non-homothetic utility function

$$u(x_1, x_2) = x_2 + [1 - (1/\varepsilon)]^{-1}\, x_1^{1-(1/\varepsilon)} + \alpha x_1 x_2 \tag{8}$$

with $\varepsilon > 0$ and $\varepsilon \neq 1$, and where $\alpha$ is close to zero- and may be either positive or negative.[4] Note that for $\alpha = 0$, this is a simple quasilinear utility function. It is the case that $u_1 = x_1^{-(1/\varepsilon)} + \alpha x_2$, $u_2 = 1 + \alpha x_1$, $u_{11} = -(1/\varepsilon)x_1^{-1/\varepsilon - 1} \leq 0$, $u_{22} = 0$. The determinant $|H|$ of the bordered Hessian $H$ equals

$$|H| = 2p_1 p_2 u_{12} - p_2^2 u_{11} - p_1^2 u_{22} = 2p_1 p_2 \alpha + p_2^2 (1/\varepsilon) x_1^{-(1/\varepsilon - 1)} \tag{9}$$

While this claim is not shown here, I argue that, for $\alpha$ close to zero, demand will also lie close to the demand for the quasilinear utility function ($\alpha = 0$), for which marginal utilities are positive, and for which $|H| > 0$. Put otherwise, as long as $\alpha$ is sufficiently close to zero, it will continue to be the case that $u_1 > 0$, $u_2 > 0$, and that $|H| > 0$ , so that the second-order condition is valid.

We now apply Cramer's rule to obtain the first derivatives of the two demand functions.

$$\frac{\partial x_1}{\partial p_1} = \frac{-x_1 p_2 u_2 + p_1 x_1 u_{22} - \lambda p_2^2}{|H|} = \frac{-p_2 x_1 (1 + \alpha x_1) - \lambda p_2^2}{|H|} \tag{10}$$

substitutability. These and other preferences on the edges of the different sets in Fig. 4 have a specific algebraic form, and are treated in [3].

[4]For $\epsilon < 1$, $u$ is not defined in $(0, x_2)$. This problem may be overcome by taking $g = e^u$ instead of $u$, with $g(0, x_2) = 0$.

$$\frac{\partial x_2}{\partial p_1} = \frac{\lambda p_1 p_2 - x_1(-p_2 u_{11} + p_1 u_{21})}{|H|} = \frac{\lambda p_1 p_2 - x_1 \left[ p_2 (1/\varepsilon) x_1^{-1/\varepsilon - 1} + p_1 \alpha \right]}{|H|} \tag{11}$$

$$\frac{\partial x_1}{\partial p_2} = \frac{\lambda p_1 p_2 + x_2(-p_1 u_{12} + p_2 u_{22})}{|H|} = \frac{\lambda p_1 p_2 + x_2(-p_1 \alpha)}{|H|} > 0 \tag{12}$$

$$\begin{aligned} \frac{\partial x_2}{\partial p_2} &= \frac{-x_2 p_1 u_1 + p_2 x_2 u_{11} - \lambda p_1^2}{|H|} \\ &= \frac{-p_1(x_1^{-(1/\varepsilon)} + \alpha x_2) + p_2 x_2(-(1/\varepsilon) x_1^{-(1/\varepsilon - 1)})}{|H|} \end{aligned} \tag{13}$$

$$\frac{\partial x_1}{\partial m} = \frac{p_2 u_{12} - p_1 u_{22}}{|H|} = \frac{p_2 \alpha}{|H|} \tag{14}$$

$$\frac{\partial x_2}{\partial m} = \frac{p_1 u_{21} - p_2 u_{11}}{|H|} = \frac{p_1 \alpha - p_2(-(1/\varepsilon) x_1^{-1/\varepsilon - 1})}{|H|} \tag{15}$$

It immediately follows that for sufficiently small $\alpha$, $\partial x_1/\partial p_1 < 0$, $\partial x_1/\partial p_2 > 0$, $\partial x_2/\partial p_2 < 0$, $\partial x_2/\partial m > 0$. Further, $\text{sgn}(\partial x_1/\partial m) = \text{sgn}(\alpha)$. Using the fact that, by the first-order condition for $x_1$, $u_1 = \lambda p_1$, it follows that

$$\frac{\partial x_2}{\partial p_1} = \frac{x_1^{-(1/\varepsilon)} p_2 \left[1 - (1/\varepsilon)\right] + \alpha(p_2 x_2 - p_1 x_1)}{|H|} \tag{16}$$

Thus, for sufficiently small $\alpha$, it is the case that $\text{sgn}(\partial x_2/\partial p_1) = \text{sgn}\left[1 - (1/\varepsilon)\right]$ (where it is easily checked that for $\alpha = 0$, $\varepsilon$ is the absolute value of the own-price elasticity of the demand for good 1). The following results are obtained for the given utility function, suggesting the existence of the preferences summarised in Fig. 4:

(i) In the case where $\varepsilon > 1$, and $\alpha$ close to zero, $\alpha > 0$, preferences are characterised by weak symmetric gross substitutability and weak symmetric income elasticity.
(ii) In the case where $\varepsilon < 1$, for $\alpha > 0$ and $\alpha$ close to zero, preferences are characterised by strong asymmetric gross substitutability, but not by strong asymmetric income elasticity.
(iii) In the case where $\varepsilon > 1$, for $\alpha < 0$ and $\alpha$ close to zero, preferences are characterised by strong asymmetric income elasticity, but not by strong asymmetric gross substitutability.
(iv) In the case where $\varepsilon < 1$, for $\alpha < 0$ and $\alpha$ close to zero, preferences are characterised by strong asymmetric gross substitutability and strong symmetric income elasticity, but there is no Giffen behaviour.

## 4 Discussion

Whether or not Giffen behaviour is actually observed empirically, insight into the determinants of Giffen behaviour is useful to gain insight into the behaviour of rational consumers, as such consumers may still approach Giffen behaviour without having upward sloping demand curves. This paper has shown that, on top of the traditional condition of inferiority of the potential Giffen good, the Giffen good should additionally be strongly asymmetrically gross substitutable for the other good, in that the Giffen good is a gross substitute for the other good, but the other good is a gross complement to the Giffen good.

To see how strong asymmetric gross substitutability increases our insight into Giffen behaviour, consider the following argument (see also [4]). In many textbooks, inferior goods are interpreted as low-quality goods, where Giffen goods are then argued to be a subclass of such low-quality goods. Yet, if a consumer consumes a low-quality good and a high-quality good, the low-quality good is a bad substitute for the high-quality good, whereas the high-quality good is a good substitute for the low-quality good. Thus, if the price of the low-quality good increases, the consumer buys a lot more of the high-quality good, but when the price of the high-quality good increases, the consumer does not buy that much more of the low-quality good. This suggests that low-quality, inferior, good is price elastic, and the high-quality good is price inelastic. But this is exactly the opposite of what we need to approach Giffen behaviour.

The necessary condition of strong asymmetric gross substitutability on the contrary suggests considering the potential Giffen good as a *high*-quality good. For instance, let a consumer consume a staple food such as rice, and another good, such as meat. Rice is a high-quality good in the sense that it is good at ensuring the subsistence of the consumer, when the consumer's subsistence is at risk. From the perspective of subsistence, rice is a good substitute for meat, whereas meat is not a good substitute for rice. From this perspective, meat is the low-quality good. At the same time, meat is a complement to rice, as rice combined with meat makes meals tasty. When the price of rice decreases, and subsistence becomes less problematic, consumption of meat may therefore increase. Similarly, when income increases, subsistence is no longer at risk, and the consumer consumes less rice.[5]

The importance of the (a)symmetry of gross substitutability suggests that insight can be increased by not only focussing on the traditional Hicksian or Slutsky decomposition, but by also developing alternative decompositions of the price effect. Hicksian decomposition may be seen as first looking at what demand would be predicted to be after a price increase if preferences were quasilinear (substitution effect), after which we look at the extent to which true demand deviates from the

[5]For an interpretation of this argument in terms of a subsistence constraint, see [5]. Another way to approach this argument is by using the characteristics approach of [8].

predicted demand (income effect). In the same manner, one could e.g. first look at what demand would be predicted to be after a price increase if preferences were homothetic, after which we can look at the extent to which true demand deviates from this. This is then the effect of weak asymmetric gross substitutability. Such alternative decompositions are the subject of future research.

## References


1. Battalio, R.C., Kagel, J.H. and Kogut, C.A. (1991). Experimental Confirmation of the Existence of a Giffen Good. American Economic Review 81:961–970.
2. Carter, M. (1995) An Expository Note on the Composite Commodity Theorem. Economic Theory 5: 175–179.
3. De Jaegher, K. (2008). Benchmark Two-Good Utility Functions. The Manchester School 76: 44–65.
4. De Jaegher, K. (2009). Asymmetric Substitutability: Theory and Some Applications. Economic Inquiry 47:838–855.
5. Gilley, O.W., and Karels, G.V. (1991). In Search of Giffen Behavior. Economic Inquiry 29: 182–189.
6. Heijman, W. and von Mouche, P.H.M. (2011). A Child Garden of Concrete Giffen Utility Functions: a Theoretical Review. 655:69–88.
7. Jensen, R.T., Miller, N.H. (2008). Giffen Behavior and Subsistence Consumption. American Economic Review 98:1553–1577.
8. Lancaster K.J. (1966). A New Approach to Consumer Theory. J Political Economy 74:132–157.

# A Child Garden of Concrete Giffen Utility Functions: A Theoretical Review

**Wim Heijman and Pierre von Mouche**

**Abstract** We present a theoretical review of the literature on concrete utility functions for Giffen and inferior goods within the context of the utility maximisation problem under a budget restriction and provide new functional forms. The presentation is organised around the specific properties such utility functions have. These properties include strict increasingness, quasi-concavity, and the applicability of Gossen's second law.

## 1 Introduction

It is generally believed that Giffen utility functions, i.e. a utility function for a Giffen good, are compatible with most qualitative properties of these functions,[1] but rigorous proofs for such beliefs generally are lacking.

The usual way to illustrate the existence of Giffen goods is by a graphical analysis for the case of two goods with a figure showing (two) indifference sets and two budget lines. This is even the case in advanced microeconomics text books like [8]. But such figures do not tell much about the specific properties the utility function may have. What we would like to have is a mathematical proof that a Giffen utility function with the desired properties exists or does not exist. One way to give an existence proof is to provide a concrete utility function with (a mathematical proof for) the desired properties. For our purposes it is sufficient to agree that a function is 'concrete' when it is a (may be complicated) composition of standard functions.

[1] An exception is homogeneity.

W. Heijman (✉) · P. von Mouche
Wageningen Universiteit, Hollandseweg 1, Postbus 8130, 6700 EW Wageningen, The Netherlands
e-mail: wim.heijman@wur.nl; pvmouche@yahoo.fr

W. Heijman and P. Mouche (eds.), *New Insights into the Theory of Giffen Goods*, Lecture Notes in Economics and Mathematical Systems 655,
DOI 10.1007/978-3-642-21777-7_6, © Springer-Verlag Berlin Heidelberg 2012

The popularity of graphical illustrations has to do with the fact that concrete Giffen utility functions with reasonable specific properties are (still) not well-known. It may be noted here that in other settings of the utility maximisation problem, for example, the labour-leisure-model it is easier to provide concrete Giffen utility functions with reasonable properties.[2]

Various authors dealt with the search for concrete Giffen utility functions $u : \mathbb{R}^2_+ \to \mathbb{R}$; the case of two goods in fact already poses enough problems.[3] So $u$ is defined on the whole $\mathbb{R}^2_+$ and not just on some subset of $\mathbb{R}^2_+$.[4] In the rest of our article we only will deal with this setting.

Depending on the specific reasonable properties such a function should have, we like to distinguish between three types of problems. We refer to these problems as 'the text book Giffen problem', the 'weak Giffen problem' and the 'strong Giffen problem'; see Sects. 3–5 for what these problems are about. Before proceeding, we emphasize that there is no universally adopted definition of a Giffen good. Definitions may differ in the details and may not be completely equivalent (also see Sect. 2).

The weak Giffen problem was solved in [19]. This was overlooked by other authors ([9,15]), who also tried to solve this problem. In fact [15] provides a solution of the text book Giffen problem and conjectured that the weak Giffen problem does not have a solution. In [9] this conjecture was refuted by providing a (technical) construction that shows that a solution does exist, without providing a concrete utility function. The idea of this proof consists in taking a backward-bending income expansion path to which a string of nearly-kinked indifference curves is attached. A whole class of utility functions that solve the weak Giffen problem was given in [13]. The strong Giffen problem recently has been solved in [2, 12].[5]

Below we review the above mentioned concrete Giffen utility functions.[6] According to economic theory there is a relationship between Giffen goods and inferior goods: under weak conditions each Giffen good is an inferior good.[7] In the contribution [11] in this book it has even been shown that

$$\text{Each upper semi-continuous Giffen utility function is inferior.} \quad (1)$$

We also will review concrete utility functions for inferior goods. However, we only are aware of a function in [6] for an inferior good that is not Giffen. Finally,

---

[2]For example in the labour-leisure-model with a Leontief utility function $u(l, x) = \min(l, x)$ and budget restriction $wl + px \leq wT$, the demand function for leisure is strictly increasing.

[3]See the contribution [14] in this book for a specific problem for the case of $n$ goods.

[4]In the case the domain is a subset of $\mathbb{R}^2_+$ it is easier to provide Giffen utility functions with reasonable properties. (Also see [3].)

[5]Doi et al. [2] is reprinted in this book.

[6]Also we mention here [17] dealing with a Giffen utility function on $\mathbb{R}_{++} \times \mathbb{R}_+$ (for which there does not exist an 'obvious counterpart' on $\mathbb{R}^2_+$). Also [17] did not seem to be aware of [19].

[7]Also see the contribution [5] in this book for other relationships for Giffen goods.

we provide a new example that easily solves the strong Giffen problem with an arbitrary prescribed precision.

We decided to make the review partially self-contained by providing in the appendix (new more modern) proofs for the correctness of various examples.[8]

## 2 Setting

The setting for the utility maximisation problems we will deal with always is the following standard one: by a 'utility function' we will understand a function $u : \mathbb{R}^2_+ \to \mathbb{R}$. And given a utility function $u$, (budget) $m \geq 0$ and (prices) $\mathbf{p} = (p_1, p_2)$ with $p_1 > 0$ and $p_2 > 0$, we consider the maximisation problem of $u$ under the (budget) restriction $p_1 x_1 + p_2 x_2 \leq m$. If $u$ is upper semi-continuous, then each utility maximisation problem has at least one maximiser. Sufficient for that there is at most one maximiser is that $u$ is strictly quasi-concave.

Given (a good) $i$, a budget and prices of the other goods, let $I$ be the subset of prices $p_i$ of good $i$ for which the utility maximisation with this budget and these prices has a unique maximiser $\check{\mathbf{x}}(p_i)$. By the 'demand function for good $i$' we understand the $i$th component function $\check{x}_i : I \to \mathbb{R}$. And given prices $\mathbf{p}$, let $J$ be the subset of budgets $m$ for which the utility maximisation with this budget and these prices has a unique maximiser $\hat{\mathbf{x}}(m)$. By the 'Engel function for good $i$' we understand the $i$th component function $\hat{x}_i : J \to \mathbb{R}$.

The definition for good $i$ to be Giffen (inferior) we will deal with is Definition 1 (Definition 2) in the contribution [11] in this book. In this definition the whole demand correspondence (Engel correspondence) matters and not only the demand function $\check{x}_i$ (Engel function $\hat{x}_i$). However, in the examples where we have to show that a utility function is Giffen, the complication that relevant utility maximisation problems have more than one maximiser will not occur: we show that good 1 is Giffen by giving a subset $\mathscr{P}$ of $I$, containing at least two elements, on which the demand function $\check{x}_1$ is strictly increasing.

We note that a utility maximisation problem with merely budget-balanced maximisers, *i.e.* maximisers $\mathbf{x}$ that satisfy $\mathbf{p} \cdot \mathbf{x} = m$, essentially comes down to maximising the function of one variable $U : [0, m/p_1] \to \mathbb{R}$ defined by

$$U(x_1) := u\left(x_1, \frac{m - p_1 x_1}{p_2}\right). \tag{2}$$

Given a utility function $u : \mathbb{R}^2_+ \to \mathbb{R}$, prices $(p_1, p_2)$ and a positive budget $m$, consider the following system of 2 equations in the 2 unknowns $x_1, x_2 \in \mathbb{R}_{++}$[9]:

[8]In doing so we correct in particular fallacies in [6, 15].

[9]The interpretation of (4) is as follows: it only is present for $(x_1, x_2) \in \mathbb{R}^2_{++}$ in which $u$ is partially differentiable.

$$p_1 x_1 + p_2 x_2 = m, \tag{3}$$

$$\frac{\partial u}{\partial x_2}(x_1, x_2)/p_2 = \frac{\partial u}{\partial x_1}(x_1, x_2)/p_1. \tag{4}$$

Equation (3) is the budget-balance restriction and the equation in (4) is also sometimes referred to as 'Gossen's second law'.

A simple result in microeconomic theory is that under weak conditions, a maximiser of the utility maximisation problem in which $u$ is partially differentiable solves (3) and (4). Indeed, each maximiser, that is interior and in which $u$ is differentiable, solves (4) and thus is a solution of the system if it is also budget balanced. A less simple, but well-known, result[10] is that sufficient conditions for a solution $(x_1, x_2)$ of the system to be a maximiser are:

*$u$ is continuous, quasi-concave and in $(x_1, x_2)$ differentiable with positive partial derivatives (Utility Maximisation Rule).*

Of course, it still might be possible that a solution $(x_1, x_2)$ of (3) and (4) is a maximiser without that the Utility maximisation Rule applies.

## 3 Text Book Giffen Problem

By the 'text book Giffen problem' we understand the following problem: give a concrete utility function $u : \mathbb{R}^2_+ \to \mathbb{R}$, a price section $\mathscr{P}$ of good 1 (i.e. a non-degenerate interval $\mathscr{P}$ of $\mathbb{R}$ with $\mathscr{P} \subseteq \mathbb{R}_{++}$) and $p_2, m > 0$ such that

- for each $p_1 \in \mathscr{P}$ the maximisation problem has a unique maximiser;
- good 1 is Giffen on $\mathscr{P}$;
- $u$ is continuous, and the restriction $u \restriction \mathbb{R}^2_{++}$ is partially differentiable;
- for each $p_1 \in \mathscr{P}$ the unique maximiser of the maximisation problem is the unique solution of the system of equations (3) and (4).

The reason that partial differentiability is assumed everywhere (in the interior of the domain of $u$) is that $u$ should be appropriate for text book purposes.[11] Without this differentiability condition a utility function in [19] (see Example 4) already solved the text book Giffen problem.

In [15] the text book Giffen problem was solved by the function $u_1^T$ in the following example (also see [18]).

[10] See for example [1].

[11] In this context, one could wish to add here 'and that this system easily can be solved by hand', because if it is difficult to solve the system of equations, the utility function is not really appropriate for text book purposes.

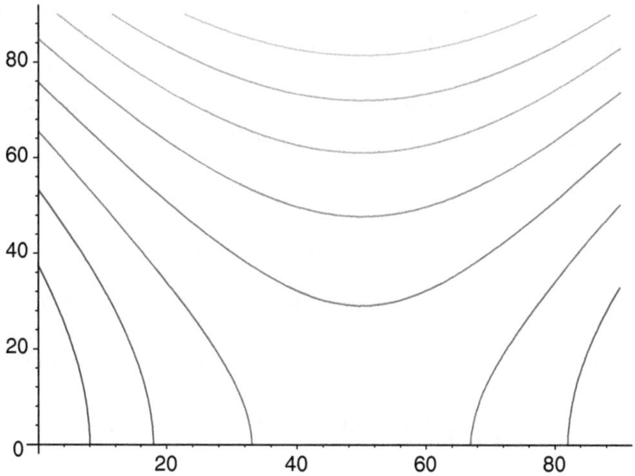


**Fig. 1** Indifference sets of $u_1^T$

*Example 1.* Let

$$u_1^T(x_1, x_2) := -x_1^2 + \alpha x_1 + x_2 + \frac{x_2^2}{2},$$

where $\alpha = 100$. $u_1^T$ solves the text book Giffen problem. More precisely: fix $p_2 = 1$ and $m \in ]1, \frac{\alpha}{\sqrt{2}} - 1[$. For $p_1 \in \mathscr{P} = ]\frac{\alpha}{m+1} - \sqrt{\frac{\alpha^2}{(m+1)^2} - 2}, \frac{\alpha}{2} - \frac{1}{2}\sqrt{\alpha^2 - 8m}[$, the utility maximisation problem has a unique maximiser.[12] This maximiser is budget-balanced, interior, and given by

$$\breve{\mathbf{x}}(p_1) = \left( \frac{-(m+1)p_1 + \alpha}{2 - p_1^2}, \frac{p_1^2 - \alpha p_1 + 2m}{2 - p_1^2} \right).$$

Good 1 is Giffen on the price section $\mathscr{P}$. ◇

Some indifference sets of $u_1^T$, *i.e.* level sets, are given in Fig. 1.[13]

It may be interesting to note that, as Example 1 shows, even additive separable functions may be Giffen.

Figure 1 (where $\mathscr{P} = \{1, 11/10\}$, $p_2 = 1$ and $m = 55$, the values that were used in [15]) shows[14] that $u_1^T$ is Giffen (by providing the two relevant indifference sets and budget lines). For this example, the analysis is easy, but proving our result

[12]Note that for small $m$ a maximiser may be not unique, as can be seen from Fig. 1 by looking to its south west corner.

[13]In all our figures the $x_1$ is on the abscissa and $x_2$ on the ordinate.

[14]Please note the scaling-down on the abscissa.

is more complicated because we have to prove that $\breve{x}_1$ is strictly increasing on the whole $\mathscr{P}$ instead of only on a two points set.

$u_1^T$ does not have so nice indifference sets as can be seen from Fig. 1.[15] After a long time of puzzling we found the following, like $u_1^T$ also quadratic, utility function that also solves the text book Giffen problem and that is, as we will explain, superior to $u_1^T$:

*Example 2.* Let

$$u_2^T(x_1, x_2) := x_2(x_2 + 4x_1 - a),$$

where $a > 0$. $u_2^T$ solves the text book Giffen problem. More precisely: fix $p_2 = 1$ and $m \in ]0, a[$. For $p_1 < \frac{4m}{a}$, the utility maximisation problem has a unique maximiser. This maximiser is budget-balanced, interior, and given by:

$$\tilde{\mathbf{x}}(p_1) = \left( \frac{4m - (2m - a)p_1}{2p_1(4 - p_1)}, \frac{4m - ap_1}{2(4 - p_1)} \right).$$

If $m > \frac{a}{2}$, then good 1 is Giffen on the price section $\mathscr{P} = ]\frac{4m-4\sqrt{am-m^2}}{2m-a}, \frac{4m}{a}[$. ◇

Figure 2 gives some indifference sets of $u_2^T$ (for $a = 800$) and illustrates for this case that $u_2^T$ is Giffen for the situation $m = 600$ and $\mathscr{P} = \{27/10, 29/10\}$.

Understanding by a 'good region' of a utility function, a non-empty convex subset of $\mathbb{R}^2_+$ with positive measure (may be infinity) where the function is strictly quasi-concave and strictly increasing, $u_2^T$ is superior to $u_1^T$ in the sense that it has (as already may be clear from the left panel in Fig. 2), contrary to $u_1^T$ a good region of a very simple form:

$$D := \{(x_1, x_2) \in \mathbb{R}^2_+ \mid x_2 > 0 \text{ and } x_2 + 4x_1 - a > 0\}. \tag{5}$$

The next example deals with a utility function in [16]:

*Example 3.* Let

$$u_3^T(x_1, x_2) := e^{\alpha x_1} x_1 e^{\frac{x_2^2}{2}},$$

where $\alpha = -\frac{1}{10}$. Fix $p_2 = 6$ and $m = 43/5$. Let, for $p_1 > 0$,

$$y_\pm(p_1) := \frac{mp_1 - p_2^2\alpha \pm \sqrt{(mp_1 - p_2^2\alpha)^2 - 4p_1^2p_2^2}}{2p_1} \in \mathbb{C}.$$

[15]Therefore in [15] also a 'repaired' version of $u_1^T$ was used to improve the properties of the indifference sets. The repair consisted on modifying $u_1^T$ for $x_1 > 50$. But this only improved these properties a little bit (the modified function still is not quasi-concave) and lead to a piecewise definition, which made this function more complicated and thereby destroyed the partially differentiability of the utility function (on $\mathbb{R}^2_{++}$).

For $p_1 \in \mathscr{P} = \{\frac{18}{20}, \frac{19}{20}\}$, the utility maximisation problem has $(\frac{y_-(p_1)}{p_1}, \frac{m-y_-(p_1)}{p_2})$ as unique maximiser. This maximiser is budget-balanced and interior. Good 1 is Giffen on the price section $\mathscr{P}$. ◇

Figure 3 clearly shows that $u_3^T$ is not quasi-concave. So $u_3^T$ neither solves the weak nor the strong Giffen problem. Because $u_3^T$ is for $p_1 \in \mathscr{P}$ differentiable in $(y_-(p_1)/p_1, (m - y_-(p_1))/p_2)$ and these maximisers are budget-balanced and interior, they are solutions of the system of equations (3)–(4). However, they are not the unique solutions: solving, for $p_1 \in \mathscr{P}$, these equations in the unknowns $x_1, x_2 \in \mathbb{C}$ gives another solution: $(y_+(p_1)/p_1, (m - y_+(p_1))/p_1)$. In fact, for $p_1 = 19/20$ we have $y_+(p_1)/p_1 = 8.142.. < 9.052.. = m/p_1$ and for $p_1 = 18/20$ we have $y_+(p_1)/p_1 = 9.134.. < 9.555.. = m/p_1$. Therefore, this particular example does not solve the Giffen text book problem. However, this does not imply that $u_3^T$

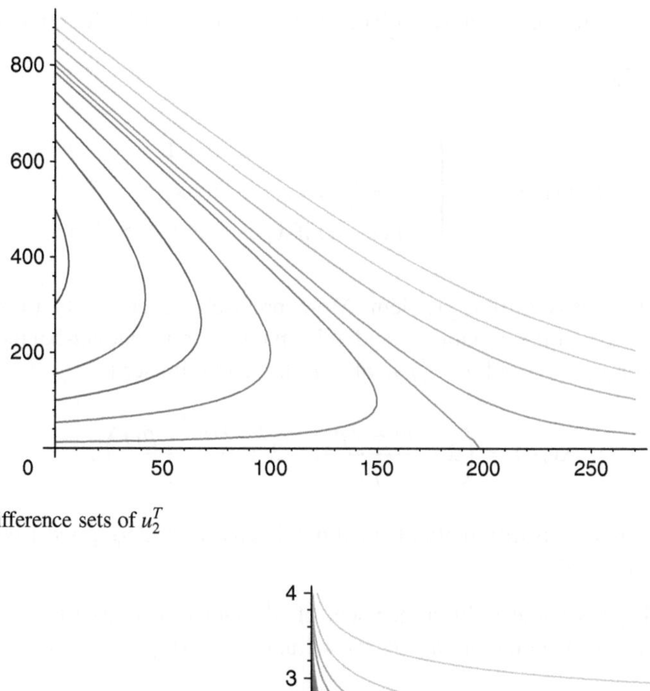


**Fig. 2** Indifference sets of $u_2^T$

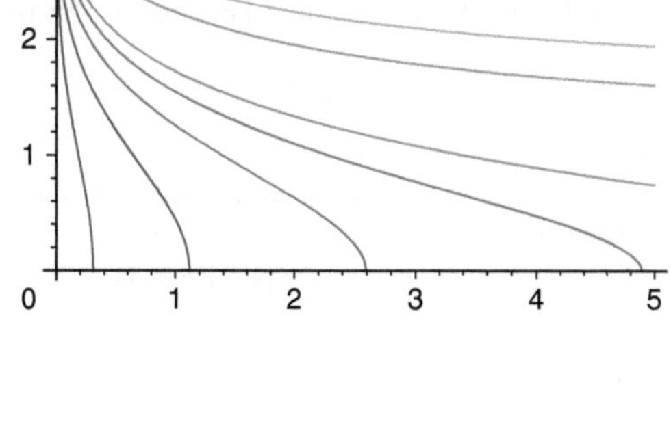


**Fig. 3** Indifference sets of $u_3^T$

does not solve this problem, may be it does for other parameter values of $\alpha, p_2, m$. Computer calculations show that for the above parameter values also other $\mathscr{P}$ do not lead to success. The above parameter values of $\alpha$, $p_2$ and $m$ are as in [16]. Thus, with these parameter values the text book Giffen problem can not be solved.[16]

## 4 Weak Giffen Problem

By the 'weak Giffen problem' we understand the following problem: give a concrete utility function $u : \mathbb{R}^2_+ \to \mathbb{R}$, a price section $\mathscr{P}$ of good 1 and $p_2, m > 0$ such that

- good 1 is Giffen on $\mathscr{P}$;
- $u$ is continuous, strictly increasing and quasi-concave.

The weak Giffen problem was solved in [19] with the utility function $u_1^W$:

*Example 4.* Let

$$u_1^W(x_1, x_2) := \begin{cases} x_1 - 1 \text{ if } x_1 \leq 1, \\ \frac{x_1-1}{(x_2-2)^2} \text{ if } x_1 > 1 \text{ and } x_2 \leq 8/5, \\ \frac{25}{4}(x_1 - 1) \text{ if } x_1 > 1 \text{ and } x_2 > 8/5. \end{cases}$$

$u_1^W$ solves the weak Giffen problem. More precisely: each utility maximisation problem has a unique maximiser and this maximiser is budget-balanced.[17] Let $m > \frac{9}{5}p_2$. For $p_1 \in \mathscr{P} = ]m - \frac{9}{5}p_2,\ m - p_2\,[$, the maximiser is

$$\check{\mathbf{x}}(p_1) = \left(2 - \frac{m - 2p_2}{p_1},\ -2 + \frac{2(m - p_1)}{p_2}\right),$$

thus interior and $u^W$ is differentiable in $\check{\mathbf{x}}(p_1)$. In case $m > 2p_2$, good 1 is Giffen on the price section $\mathscr{P}$. ◇

Figure 4 gives some indifference sets of $u_1^W$ and illustrates for this case that $u_1^W$ is Giffen for the situation $m = 4$, $p_2 = 1$ and (the two points set) $\mathscr{P} = \{45/20, 48/20\}$.

Analysing the with $u_1^W$ associated function $U$ (given by (2)) here is straightforward but a little technical because of the piece-wise definition of $U$. The more

[16]Even if it can with other parameter values, then it is correct to say that this problem first was solved in [15].

[17]But not necessarily interior for positive budget.

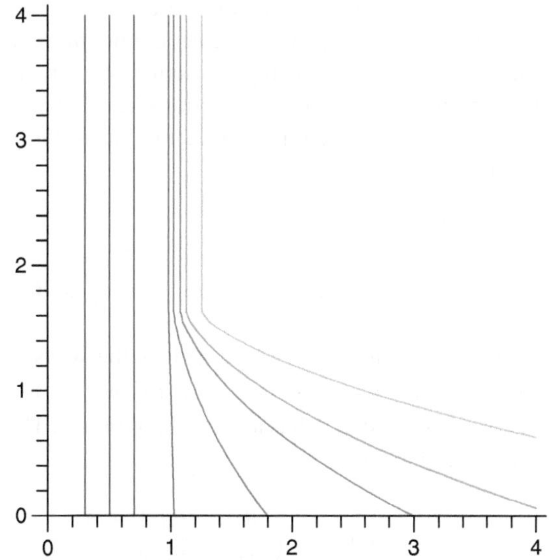


**Fig. 4** Indifference sets of $u_1^W$

efficient proof we give (in the appendix) uses some theoretical arguments and the Utility Maximisation Rule from Sect. 2.

Note that in Example 4, the demand function $\breve{x}_1$ is strictly concave on $\mathscr{P}$. This is very nice, because demand functions that are strictly concave on some non-empty interval seem to be rare.[18] Also nice is that the demand function $\breve{x}_1$ even is well-defined for all prices $p_1$. The reader may check that in case $m = 4$ and $p_2 = 1$ (by analysing $U$ outside $\mathscr{P}$) that it is given by

$$\breve{x}_1(p_1) = \begin{cases} \frac{12}{5p_1}\ (p_1 \leq 11/5), \\ \frac{2p_1-2}{p_1}\ (11/5 < p_1 < 3), \\ \frac{m}{p_1}\ (p_1 \geq 3). \end{cases}$$

As our proof of Example 4 shows, in this example even all conditions for the text book Giffen problem are satisfied, with the exception of the partial differentiability on $\mathbb{R}^2_{++}$. The reason for this shortcoming is, like is often the case, that the utility function is piece-wise defined.

A whole class of other utility functions that also solve the weak Giffen problem was given in [13]. The interesting idea there is to use Leontief-like utility functions $u$ of the form

$$u(x_1, x_2) = \min(f_1(x_1, x_2), f_2(x_1, x_2))$$

[18]Even demand functions that are not log-convex seem to be rare.

with a widened angle at the indifference set kink. It is not so clear who first had this idea (also see [4]). Possible economic interpretations of such composed utility functions can be given using the notion of characteristics of goods; see [7]. Due to the min-function, the utility functions here are also not partially differentiable – in fact the piece-wise definition aspect remains. We present here only one element of the class:

*Example 5.* Let

$$u_2^W(x_1, x_2) := \min(x_2 + B,\ A(x_1 + x_2)),$$

where $A > 1$ and $B > 0$. $u_2^W$ solves the weak Giffen problem. More precisely: suppose $0 < \frac{m(A-1)}{B} < p_2$. Then for $p_1 \in \mathscr{P} = ]0, \min(m\frac{A}{B}, p_2)[$, the utility maximisation problem has a unique maximiser. This maximiser is budget-balanced, interior, and given by

$$\check{\mathbf{x}}(p_1) = \left( \frac{Bp_2 - m(A-1)}{Ap_2 - p_1(A-1)}, \frac{Am - Bp_1}{Ap_2 - p_1(A-1)} \right);$$

$u_2^W$ is not differentiable in $\check{\mathbf{x}}(p_1)$. Good 1 is Giffen on the price section $\mathscr{P}$. ◇

Although $u_2^W$ is less complicated than $u_1^W$, the latter has the advantage that each utility maximisation problem has a unique maximiser. Figure 5 gives some indifference sets of $u_2^W$ for $A = 2$ and $B = 10$ and illustrates for this case that $u_2^W$ is Giffen for the situation $m = 5$, $p_2 = 1$ and (the two points set) $\mathscr{P} = \{1/3, 3/4\}$.

In Sect. 3 we have seen that the text book Giffen problem already can be solved with a quadratic utility function. However, we do not know whether there is a quadratic utility function that solves the weak Giffen problem.

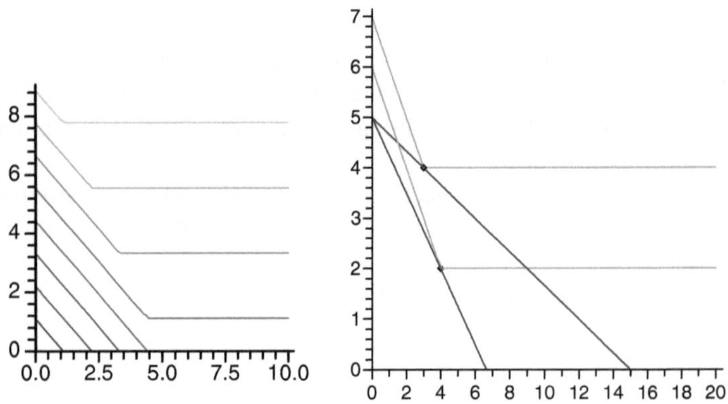


**Fig. 5** Indifference sets of $u_2^W$ (*left*), and $u_2^W$ is Giffen (*right*)

## 5 Strong Giffen Problem

By the "strong Giffen problem" we understand the following problem: give a concrete utility function $u : \mathbb{R}^2_+ \to \mathbb{R}$ that simultaneously solves the weak and text book Giffen problems.

The strong Giffen problem recently has been solved in [2]:

*Example 6.* The utility function

$$u_1^S(x_1, x_2) := \begin{cases} x_1^{\alpha_1} x_2^{\alpha_2} e^{-\gamma x_1 x_2} & \text{if } x_1 x_2 < \alpha/\gamma, \\ (\alpha/\gamma)^{\alpha} x_2^{\beta-\alpha} e^{-\alpha} & \text{otherwise,} \end{cases}$$

where $\beta > \alpha > \beta/2 > 0$ and $\gamma > 0$, solves the strong Giffen problem. In addition: the utility function is differentiable on $\mathbb{R}^2_{++}$,[19] each utility maximisation problem has a unique maximiser and this maximiser is budget-balanced and, for positive budget, interior. ◇

The proof in [2] of the correctness of the statements in this example was a whole 'tour de force'. Because this proof is sound and we can not simplify it substantially, we will not give a (new) proof here.

That there exists a utility function that solves the strong Giffen problem became pretty clear from an abstract argument given in [13]: approximate the function $\min(z_1, z_2)$ in $u_2^W$ by the (quasi-concave) ces-function $(z_1^{\rho} + z_2^{\rho})^{1/\rho}$ for $\rho \to -\infty$. For solving the strong Giffen problem one can try to work out this idea. This indeed can be done with success and leads to the function $U_2^S$ in the following Example 7 taken from [12]. The example also provides another solution of the strong Giffen problem: the function $U_3^S$. The finding of $U_2$ was less straightforward and based on a different idea. In fact it was inspired by the construction in [9], mentioned in the introduction. (Also see the contribution [10] in this book.)

*Example 7.* The following two utility functions

$$U_2^S(x_1, x_2) := 100(x_1 + x_2)(x_2 + 114)\Big((100(x_1 + x_2))^4 + (x_2 + 114)^4\Big)^{-1/4},$$

$$U_3^S(x_1, x_2) := x_1 + \frac{3}{2}x_2 - \sqrt{\left(x_1 + \frac{x_2}{2} - 25\right)^2 + 5}.$$

solve the strong Giffen problem. ◇

Finally, we return to our utility function $u_2^T$ which solves the text book Giffen problem. The interesting observation is, that its definition is such that the measure

[19] But not smooth.

of the complement of its good region $D$ (given by (5)), is $a^2/8$ and that thus this measure may be made arbitrary small. The conclusion is that given an arbitrary small positive number $\epsilon$, the utility function $u_2^T$ (for $a = \sqrt{8\epsilon}$) not only solves the text book Giffen problem but also the strong Giffen problem 'modulo a bad region of its domain with measure $\epsilon$'.[20]

# 6 Inferiority Problems

We have defined three types of Giffen pròblems: the weak, text book and strong Giffen problem. By modifying the definitions of these notions in a natural way we obtain the corresponding three types of inferiority problems.

Because in each of the previous examples the utility function is upper semi-continuous, the utility functions there are inferior according to (1). Therefore $u_1^W, u_2^W$ solve the weak inferiority problem, $u_1^T, u_2^T$ solve the text book inferiority problem and $u_1^S, u_2^S, u_3^S$ solve the strong inferiority problem. Concerning (other) concrete inferior utility functions we of course have $u_3^T$. In addition we found one in [6][21]:

*Example 8.* Let

$$u_1^I(x_1, x_2) := x_1^\alpha e^{x_2^2/2},$$

where $\alpha > 0$. $u_1^I$ is inferior. More precisely: for $p_2 > 0$ and $m > 2p_2\sqrt{\alpha}$, let

$$y_\pm(m) := \frac{m \pm \sqrt{m^2 - 4p_2^2\alpha}}{2}, \quad t(m) := \left(\frac{y_-(m)}{m}\right)^\alpha e^{\frac{y_+^2(m)}{2p_2^2}} - 1.$$

- If $m \leq 2p_2\sqrt{\alpha}$, then $\hat{\mathbf{x}}(m) = (m/p_1, 0)$ is the unique maximiser of the utility maximisation problem.
- If $m > 2p_2\sqrt{\alpha}$ and $t(m) \neq 0$, then the utility maximisation problem has a unique maximiser. This maximiser is budget-balanced and given by

[20] Furthermore it may be worthwhile to note that in Example 2 the maximisers are in the good region $D$ (as the proof of the example shows).

[21] In fact in [6] the correctness of the claim that the utility function is inferior is not proved. There just is referred to a book with a not correct analysis: it is stated that the formula for $\hat{\mathbf{x}}(m)$ in Example 8 for the case $m > 2p_2\sqrt{\alpha}$ and $t(m) > 0$ also is correct in case $t(m) \leq 0$. However, for $\alpha = \frac{1}{10}$, $p_1 = 1$, $p_2 = 1$, $m = \frac{21}{10}p_2\sqrt{a}$, one has $t(m) < 0$ which (now referring to our proof in the appendix) implies $U(m/p_1) > U(y_-(m)/p_1)$ from which it follows that $(0, m/p_1)$ is the unique maximiser instead of $(y_-(m)/p_1, y_+(m)/p_2)$.

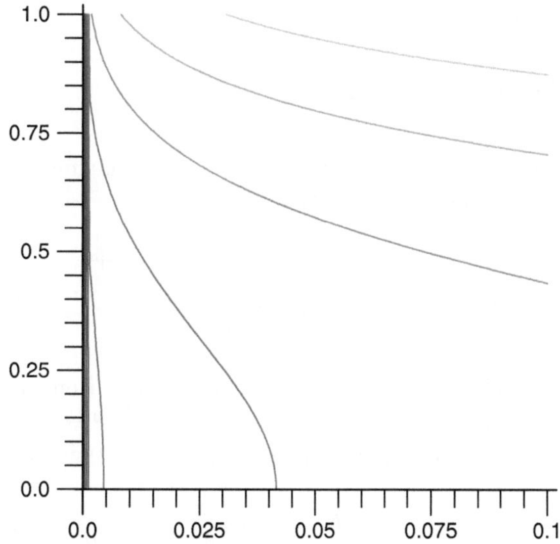


**Fig. 6** Indifference sets of $u_1^I$

$$\hat{\mathbf{x}}(m) = \begin{cases} \left(\frac{y_-(m)}{p_1}, \frac{y_+(m)}{p_2}\right) \text{ if } t(m) > 0, \\ \left(\frac{m}{p_1}, 0\right) \text{ if } t(m) < 0. \end{cases}$$

- If $m > 2p_2\sqrt{\alpha}$ and $t(m) = 0$, then $\left\{\left(\frac{y_-(m)}{p_1}, \frac{y_+(m)}{p_2}\right), \left(\frac{m}{p_1}, 0\right)\right\}$ is the set of maximisers.

Finally, given $\alpha, p_1, p_2 > 0$, there exists $m_0 > 2p_2\sqrt{\alpha}$ such that good 1 is inferior on the budget section $[m_0, \infty[$. $\diamond$

Figure 6 gives some indifference sets of $u_1^I$ in case $\alpha = 1/10$ and illustrates for this case that good 1 is inferior on $\mathcal{M} = \{1, 2\}$ (for the situation $p_1 = 1, p_2 = 1$). The figure also shows that $u_1^I$ is not quasi-concave and therefore does not solve the weak inferiority problem. In order to show that $u_1^I$ does not solve the text book inferiority problem either, we consider the system of equations (3) and (4). They only have a solution in case $m \geq 2p_2\sqrt{\alpha}$. In case $m = 2p_2\sqrt{\alpha}$, the unique solution is $(m/2p_1, m/2p_2)$, but the result in Example 8 shows that this is not a maximiser of the utility maximisation problem. And in case $m > 2p_2\sqrt{\alpha}$, there is not a unique solution, but there are two different solutions: $(y_-(m)/p_1, y_+(m)/p_2)$ and $(y_+(m)/p_1, y_-(m)/p_2)$.

Note that the formulas in Example 8 show that $u_1^I$ is not Giffen. Thus $u_1^I$ is inferior but not Giffen.

We like to make a conjecture: no additive separable and no multiplicative separable utility function solves the weak inferior problem.

# Appendix

## *Proof of Example 1*

Since $u_1^T$ is strictly increasing in $x_2$, each maximiser is budget-balanced. Fix $p_2 = 1$ and $1 \leq m \leq \frac{\alpha}{\sqrt{2}} - 1$. Now

$$c_1^{\pm}(m) := \frac{\alpha}{m+1} \pm \sqrt{\frac{\alpha^2}{(m+1)^2} - 2}, \quad c_2(m) := \frac{\alpha}{2} - \frac{1}{2}\sqrt{\alpha^2 - 8m}$$

are well defined real numbers, and for $m \neq 1$ one has that $0 < c_1^-(m)$ and $c_2(m) < \sqrt{2}$. Also one has $c_1^-(m) < c_2(m)$,[22] thus $\mathscr{P} = ]c_1^-(m), c_2(m)[$ is a price section.

Now consider the function $U$ given by (2) where $p_1 \in \mathscr{P}$:

$$U(x_1) = \left(\frac{1}{2}p_1^2 - 1\right)x_1^2 + (\alpha - (m+1)p_1)x_1 + \frac{1}{2}m^2 + m.$$

$U$ is a quadratic concave function. Considered as a function on the whole $\mathbb{R}$ it has

$$\breve{x}_1(p_1) = \frac{-(m+1)p_1 + b}{2 - p_1^2}$$

as unique maximiser. However, the inequalities $m < \frac{\alpha}{\sqrt{2}} - 1$ and $p_1 < \sqrt{2}$ imply that $\breve{x}_1(p_1) > 0$ and $p_1 < c_2(m)$ implies that $\breve{x}_1(p_1) < \frac{m}{p_1}$. Thus we can conclude that $\breve{x}_1(p_1)$ is a maximiser of $U$. Its derivative is

$$\breve{x}_1'(p_1) = -\frac{m+1}{(p_1^2-2)^2}\left(p_1^2 - \frac{2\alpha}{m+1}p_1 + 2\right)$$

$$= -\frac{m+1}{(p_1^2-2)^2}\left(p_1 - \left(\frac{\alpha}{m+1} + \sqrt{\frac{\alpha^2}{(m+1)^2} - 2}\right)\right)$$

---

[22]Here is a proof: let $f$ be the function $c_1^- - c_2$. Suppose $m$ is a zero of $f$. Then $(\alpha/(m+1) - \alpha/2)^2 = (\sqrt{\alpha^2/(m+1)^2 - 2} - \sqrt{(\alpha^2 - 8m)/2}\,)^2$. Evaluating this equation, next simplifying and then squaring the result after appropriate rearranging, leads to $2m^4 - (\alpha^2+4)m^2 + 2m\alpha^2 - \alpha^2 + 2 = 0$. Because $2m^4 - (\alpha^2+4)m^2 + 2m\alpha^2 - \alpha^2 + 2 = 2(m-1)^2(m - \alpha/\sqrt{2} - 1)(m + \alpha/\sqrt{2} + 1)$, we can conclude that the zeros of $f$ are 1 and $\alpha/\sqrt{2} - 1$. Noting that $f'(1) = (\sqrt{\alpha^2 - 8} - a) < 0$ and that $f$ is continuous, it follows that $f < 0$ on the interval $(1, \alpha/\sqrt{2} - 1)$, as desired.

It would be interesting to understand the deeper meaning why the inequality $c_1^-(m) < c_2(m)$, that in fact came up in a natural way, holds.

$$\times\left(p_1+\left(\frac{\alpha}{m+1}-\sqrt{\frac{\alpha^2}{(m+1)^2}-2}\right)\right)$$

$$=-\frac{m+1}{\left(p_1^2-2\right)^2}(p_1-c_1^+(m))(p_1-c_1^-(m))$$

and we see that $\breve{x}_1'(p_1) > 0$ on $\mathscr{P}$, as desired.

In order to see that $u_1^T$ solves the text book Giffen problem, one finally has to check simply that $\breve{\mathbf{x}}(p_1)$ is the unique solution of the system of equations (3) and (4). □

## *Proof of Example 2*

Suppose $p_2 = 1$. For the function $U$ given by (2) we have

$$U(x_1)=(m-p_1x_1)\left((m-p_1x_1)+4\left(x_1-\frac{a}{4}\right)\right)=(m-p_1x_1)((4-p_1)x_1+m-a).$$

In case $m/p_1 > a/4$, we see there is a left punctured neighbourhood of $m/p_1$ where $U(x_1)$ is positive. And in case $m > a$, we see that there is a right punctured neighbourhood of 0 where $U(x_1)$ is positive. This implies that in both these cases the maximum of the utility maximisation problem is positive. This in turn implies that in these both cases for each maximiser $(x_1^\star, x_2^\star)$ of the utility maximisation problem one has $x_2^\star > 0$ and $x_2^\star + 4x_1^\star - a > 0$ and therefore $(x_1^\star, x_2^\star) \in D$ (with $D$ given by (5)). Because on $D$ the utility function is strictly increasing, it follows that $(x_1^\star, x_2^\star)$ is budget-balanced and therefore that we can use $U$ also for further analysis in these cases.

Now suppose further that $m/p_1 > a/4$ and $m < a$. Noting that

$$U(x_1) = (p_1^2 - 4p_1)x_1^2 + (4m - 2mp_1 + p_1a)x_1 + m^2 - ma,$$

studying $U$ quickly gives that the utility maximisation problem has a unique maximiser $\breve{\mathbf{x}}(p_1)$ as desired. We have

$$\frac{d\breve{x}_1}{dp_1}(p_1)=-\left(m-\frac{a}{2}\right)\frac{\left(p_1-\frac{4m+4\sqrt{am-m^2}}{2m-a}\right)\left(p_1-\frac{4m-4\sqrt{am-m^2}}{2m-a}\right)}{p_1^2(p_1-4)^2}.$$

Now further supposing that $m > a/2$, on $\mathscr{P}$ this derivative is positive and thus good 1 is Giffen on $\mathscr{P}$.

In order to see that $u_2^T$ solves the text book Giffen problem one has to check simply that $\breve{\mathbf{x}}$ is the unique solution of the system of equations (3) and (4). □

## *Proof of Example 3*

Because $u_3^T$ is strictly increasing, each utility maximisation problem only has budget-balanced maximisers and therefore we can study the function $U : [0, m/p_1] \to \mathbb{R}$ given by (2):

$$U(x_1) = e^{\alpha x_1} x_1 e^{\frac{(m - p_1 x_1)^2}{2p_2^2}}.$$

Now let $m > 0$. It is clear that 0 is not a maximiser of $U$. $U$ is differentiable and, for $x_1 \neq 0$,

$$U'(x_1) = \frac{U(x_1)}{x_1} q(x_1),$$

where

$$q(x_1) := \frac{p_1^2}{p_2^2} \left( x_1^2 - \frac{mp_1 - p_2^2 \alpha}{p_1^2} x_1 + \frac{p_2^2}{p_1^2} \right).$$

Note that we have

$$q = \frac{p_1^2}{p_2^2} \left( x_1 - \frac{y_-(p_1)}{p_1} \right) \left( x_1 - \frac{y_+(p_1)}{p_1} \right).$$

It follows that for $x_1 \neq 0$

$$\begin{aligned} U''(x_1) = \frac{p_1^2}{p_2^2} \frac{U(x_1)}{x_1^2} \Bigg( \frac{p_1^2}{p_2^2} \left( x_1 - \frac{y_-(p_1)}{p_1} \right)^2 \left( x_1 - \frac{y_+(p_1)}{p_1} \right)^2 \\ - \left( x_1 - \frac{y_-(p_1)}{p_1} \right) \left( x_1 - \frac{y_+(p_1)}{p_1} \right) \\ + x_1 \left( x_1 - \frac{y_-(p_1)}{p_1} \right) + x_1 \left( x_1 - \frac{y_+}{p_1} \right) \Bigg). \end{aligned}$$

In case $y_-(p_1)/p_) \in \mathbb{R}$ it holds that $0 < y_-(p_1)/p_1 < y_+(p_1)/p_1$ and $U''(y_-(p_1)/p_1) < 0$, $U''(y_+(p_1)/p_1) > 0$. In addition this then implies that the non-empty set of maximisers of $U$ is a subset of $\{y_-(p_1)/p_1, m/p_1\} \cap [0, m/p_1]$. We have

$$\frac{y_-(p_1)}{p_1} \in \left[0, \frac{m}{p_1}\right] \Rightarrow U_i := U\left(\frac{y_-(p_1)}{p_1}\right) = e^{-\frac{y_-(p_1)}{10p_1}} \frac{y_-(p_1)}{p_1} e^{(\frac{43}{5} - y_-(p_1))^2/72}.$$

$$U_r := U\left(\frac{m}{p_1}\right) = e^{-\frac{43}{50p_1}} \frac{43}{5p_1}.$$

For $p_1 = 19/20$ we have $y_-(p_1)/p_1 = 4.898.. \leq 172/19 = m/p_1$, $U_i = 3.726..$ and $U_r = 3.661..$; thus $y_-(p_1)/p_1 = 4.898..$ is the unique maximiser. For $p_1 = 18/20$

we have $y_{-}(p_1)/p_1 = 4.865.. \leq 86/9 = m/p_1$, $U_i = 3.838..$ and $U_r = 3.675..$; thus $y_{-}(p_1)/p_1 = 4.865..$ is the unique maximiser. It follows that good 1 is Giffen on $\{18/20, 19/20\}$. □

## *Proof of Example 4*

It is straightforward to check that $u_1^W$ is strictly increasing and continuous. Quasi-concavity of $u_1^W$ may now be clear from the shape of the upper level sets in Fig. 4; we omit a formal proof. Because $u_1^W$ is strictly increasing each maximiser is budget-balanced. Next we prove by contradiction that each utility maximisation problem has a unique maximiser. So suppose $\mathbf{a}, \mathbf{b}$ would be two different maximisers. Because $u_1^W$ is quasi-concave, the set of maximisers is convex and therefore each point on the segment with $\mathbf{a}$ and $\mathbf{b}$ as end points would be a maximiser too. Because all these points are budget-balanced, we would have a non-vertical line segment on which $u$ is constant, which clearly is (by Fig. 4) not the case.

Next, let $m > \frac{9}{5}p_2$, and consider the utility maximisation problem for $p_1 \in \mathscr{P}$. Let $\breve{\mathbf{x}}(p_1)$ as given. Then $p_1\breve{x}_1(p_1)+p_2\breve{x}_2(p_1) = m$. Because $p_1 \in \mathscr{P}$ it follows that $0 < \breve{x}_1(p_1) < m/p_1$. Therefore $\breve{\mathbf{x}}(p_1) \in \mathbb{R}^2_{++}$ and thus $\breve{\mathbf{x}}(p_1)$ is a solution of (3). Noting that also $\breve{x}_1(p_1) > 1$ and $\breve{x}_2(p_1) < 8/5$, $u$ is in $\breve{\mathbf{x}}(p_1)$ differentiable and one easily checks that $\breve{\mathbf{x}}(p_1)$ also is a solution of (4), and even that $\frac{\partial u}{\partial x_2}/p_2 = \frac{\partial u}{\partial x_1}/p_1 > 0$. Because $u$ is quasi-concave, the Utility Maximisation Rule guarantees that $\breve{\mathbf{x}}(p_1)$ is a maximiser. If $m > 2p_2$, then $u_1^W$ is Giffen on $\mathscr{P}$. □

## *Proof of Example 5*

Of course, $u_2^W$ is continuous and strictly increasing. Therefore each utility maximisation problem only has budget-balanced maximisers. And $u_2^W$ is (quasi-)concave because it is a minimum of two concave functions.

For the utility maximisation problem we study again the function $U$ defined by (2). We have

$$U(x_1) = \min\left(-\frac{p_1}{p_2}x_1 + \frac{m}{p_2} + B,\ A\left(1 - \frac{p_1}{p_2}\right)x_1 + \frac{Am}{p_2}\right).$$

Further let $0 < m\frac{A-1}{B} < p_2$, $p_1 \in \mathscr{P} = ]0, \min(m\frac{A}{B}, p_2)[$. Now $0 < \frac{Bp_2 - m(A-1)}{Ap_2 - p_1(A-1)} < \frac{m}{p_1}$, $p_1 < p_2$ and

$$U(x_1) = \begin{cases} A(1 - \frac{p_1}{p_2})x_1 + \frac{Am}{p_2} \text{ if } 0 \leq x_1 \leq \frac{Bp_2 - m(A-1)}{Ap_2 - p_1(A-1)}, \\ -\frac{p_1}{p_2}x_1 + \frac{m}{p_2} + B \text{ if } \frac{Bp_2 - m(A-1)}{Ap_2 - p_1(A-1)} \leq x_1 \leq \frac{m}{p_1}. \end{cases}$$

We see that the unique maximiser of $U$ is $\frac{Bp_2 - m(A-1)}{Ap_2 - p_1(A-1)}$, and thus there $U$ is not differentiable. Thus good 1 is Giffen on $\mathscr{P}$. □

## *Proof of Example 8*

Because $u_1^I$ is strictly increasing, each utility maximisation problem only has budget-balanced maximisers and therefore we can study the function $U$ given by (2). We have

$$U(x_1) = x_1^{\alpha} e^{\frac{(m - p_1 x_1)^2}{2p_2^2}}.$$

Now let $m > 0$, It is clear that 0 is not a maximiser of $U$. $U$ is on $]0, m/p_1]$ differentiable and there

$$U'(x_1) = \frac{U(x_1)}{x_1} q(x_1),$$

where

$$q(x_1) := \frac{p_1^2}{p_2^2} \left( x_1^2 - \frac{m}{p_1} x_1 + \alpha \frac{p_2^2}{p_1^2} \right).$$

Note that in case $m > 2p_2\sqrt{\alpha}$, we have $0 < y_-(m)/p_1 < y_+(m)/p_1 < m/p_1$ and

$$q = \frac{p_1^2}{p_2^2} \left( x_1 - \frac{y_-(m)}{p_1} \right) \left( x_1 - \frac{y_+(m)}{p_1} \right).$$

Case $(0 <) m < 2p_2\sqrt{\alpha}$. Now $q > 0$ and therefore $U' > 0$ on $]0, m/p_1]$. Thus $U$ is strictly increasing on $]0, m/p_1]$ and therefore $m/p_1$ is the unique maximiser of $U$.

Case $m = 2p_2\sqrt{\alpha}$. Now again $U$ is strictly increasing on $]0, m/p_1]$ and thus $m/p_1$ is the unique maximiser of $U$.

Case $m > 2p_2\sqrt{\alpha}$. Now on $]0, m/p_1]$, $U'(x_1) = \frac{p_1^2}{p_2^2} \frac{U(x_1)}{x_1} (x_1 - \frac{y_-(m)}{p_1})(x_1 - \frac{y_+(m)}{p_1})$. So only $y_-(m)/p_1, y_+(m)/p_1$ and $m/p_1$ may be a maximiser of $U$. $U$ is on $]0, m/p_1]$ twice differentiable:

$$\begin{aligned} U''(x_1) = \frac{p_1^2}{p_2^2} \frac{U(x_1)}{x_1^2} \Bigg( & \frac{p_1^2}{p_2^2} \left( x_1 - \frac{y_-(m)}{p_1} \right)^2 \left( x_1 - \frac{y_+(m)}{p_1} \right)^2 \\ & - \left( x_1 - \frac{y_-(m)}{p_1} \right) \left( x_1 - \frac{y_+(m)}{p_1} \right) \\ & + x_1 \left( x_1 - \frac{y_-(m)}{p_1} \right) + x_1 \left( x_1 - \frac{y_+(m)}{p_1} \right) \Bigg), \end{aligned}$$

it follows that $U''(y_-(m)/p_1) < 0$ and $U''(y_+(m)/p_1) > 0$. This implies that only $y_-(m)/p_1$ and $m/p_1$ may be a maximiser of $U$. Now we compare $U(y_-(m)/p_1)$ and $U(m/p_1)$. The condition $t(m) > 0$ is equivalent with $U(y_-(m)/p_1) > U(m/p_1)$ and therefore in this case $y_-(m)/p_1$ is the unique maximiser. The condition $t(m) < 0$ is equivalent with $U(y_-(m)/p_1) < U(m/p_1)$ and therefore in this case $m/p_1$ is the unique maximiser. The condition $t(m) = 0$ is equivalent with $U(y_-(m)/p_1) = U(m/p_1)$ and therefore in this case the set of maximisers is $\{y_-(m)/p_1, m/p_1\}$.

In order to prove the last statement we note that $\frac{dy_-}{dm} < 0$ and analyse the function $t : ]2p_2\sqrt{\alpha}, \infty[ \to \mathbb{R}$. We have

$$t = \left(\frac{1}{2}\left(1 - \sqrt{1 - \frac{4\alpha p_2^2}{m^2}}\right)\right)^{\alpha} e^{\frac{\left(m+\sqrt{m^2-4p_2^2\alpha}\right)^2}{8p_2^2}} - 1.$$

This implies that $\lim_{t\to\infty} t(m) = \infty$. So there exist $m_0 > 2p_2\sqrt{\alpha}$ such that $t(m) > 0$ $(m \geq m_0)$. By the above good 1 is inferior on $[m_0, \infty[$.[23] □

## References

1. E. Balder. Exact and useful optimization methods for microeconomics. 655:19–35, 2011.
2. J. Doi, K. Iwasa, and K. Shimomura. Giffen behavior independent of the wealth level. *Economic Theory*, 41:247–267, 2009 and 655:99–118, 2011.
3. R. Haagsma. Notes on some theories of giffen behaviour. 655:5–18, 2011.
4. K. De Jaegher. Asymmetric substitutability: Theory and some applications. Economic Inquiry, 47(4):838-855, 2009.
5. K. de Jaegher. Giffen behaviour and strong assymmetric gross substitutability. 655:51–34, 2011.
6. H. Liebhafsky. New thoughts about inferior goods. *American Economic Review*, 59:931–934, 1969.
7. R. Lipsey and G. Rosenbluth. A contribution to the new theory of demand: a rehabilitation of the Giffen good. *Revue canadienne d'Economique*, IV(2):131–163, 1971.
8. A. Mas-Colell, M. D. Whinston, and J. R. Green. *Microeconomic Theory*. Oxford University Press, New York, 1995.
9. P. Moffatt. Is Giffen behaviour compatible with the axioms of consumer theory? *Journal of Mathematical Economics*, 37:259–267, 2002.

[23] A straightforward calculation shows that

$$t' = \frac{1}{4}\frac{\left(\sqrt{m^2-4p_2^2\alpha}\left(m+\sqrt{m^2-4p_2^2\alpha}\right)\right)e^{\frac{\left(m+\sqrt{m^2-4p_2^2\alpha}\right)^2}{8p_2^2}}\left(m+\sqrt{m^2-4p_2^2\alpha}\right)\left(\frac{m-\sqrt{m^2-4p_2^2\alpha}}{2m}\right)^{\alpha}}{mp_2^2\sqrt{m^2-4p_2^2\alpha}}.$$

This implies that $t' > 0$ and thus that $t$ is strictly increasing. Therefore for $m_0$ even each $m_0 > 2p_2\sqrt{\alpha}$ with $t(m_0) > 0$ can be taken.

10. P. Moffatt. A Class of Indirect Utility Functions Predicting Giffen Behaviour. 655: 119–129 2011.
11. P. H. M. von Mouche and W. Pijnappel. On the definitions of Giffen and inferior goods. 655:39–52, 2011.
12. P. G. Moffatt, P. H. M. von Mouche, and E. Smits. On concrete Giffen utility functions with very nice theoretical properties. In preparation, 2011.
13. P. Sørenson. Simple utility functions with Giffen demand. *Economic Theory*, 31:367–70, 2006.
14. P. Sørenson. Giffen demand for several goods. 655:97–104, 2011.
15. U. Spiegel. The case of a 'Giffen good'. *Journal of Economic Education*, 25:137–147, 1994.
16. E. Silberberg and D. Walker. A modern analysis of Giffen's paradox. *International Economic Review*, 25(3):687–694, 1984.
17. D. Vandermeulen. Upward sloping demand curves without the Giffen paradox. *American Economic Review*, 2:453–458, 1972.
18. C. Weber. The case of a Giffen good: Comment. *Journal of Economic Education*, 28:36–44, 1997.
19. H. Wold. On Giffen's paradox. *Nordisk Tidsskrift for Teknisk Økonomi*, 12:283–290, 1948.

# On Giffen's Paradox

**Herman O.A. Wold**

Comment from the editors: this article concerns a reprint from *Nordisk Tidsskrift for Teknisk Økonomi*, 12:283–290, 1948. As there were no abstract and no section headings we have created them ourselves.

**Abstract** A continuous, strictly increasing quasi-concave utility function that exhibits Giffen behaviour is given.

## 1 Giffen's Paradox

The phenomenon known as Giffen's paradox consists in an exception from the general rule that rising (falling) price brings, ceteris paribus, falling (rising) consumption. Interpreting the reverse behaviour of the consumer by means of an indifference map,[1] the present note gives an explicit example of Giffen's paradox. The map has been worked out for teaching purposes primarily, as no example of this kind has been set forth earlier, as far as I know. Now since Giffen's phenomenon is rather peculiar, it is no wonder that the indifference map presented is rather peculiar too. The fact that we are dealing with an exceptional situation is further brought out by a theorem to the effect that Giffen's phenomenon is necessarily restricted to a limited sector of the indifference map.

[1]For the notion of indifference map and related concepts employed in this paper, see e. g. H. Wold: A synthesis of pure demand analysis I–III, Skandinavisk Aktuarietidsskrift 1943–1944.

H.O.A. Wold (✉)
(† 1992)
Uppsala University, Sweden

W. Heijman and P. Mouche (eds.), *New Insights into the Theory of Giffen Goods*, Lecture Notes in Economics and Mathematical Systems 655,
DOI 10.1007/978-3-642-21777-7_7, 

## 2 Indifference Map

Our indifference map is shown in Fig. 1. We shall speak of the map as referring to an individual consumer. The coordinates of the map, $x$ and $y$, represent quantities of two commodities, say $\xi$ and $\eta$. The map is formed by two regions, say $\Phi$ and $\Psi$. We define $\Phi$ and $\Psi$ to consist of those points $(x,\ y)$ for which

$$x > 1 \quad \text{and} \quad 0 \leq y \leq 1 \cdot 6 \tag{$\Phi$}$$

$$\text{either } 0 \leq x \leq 1 \quad \text{or } x > 1 \text{ and } y \geq 1 \cdot 6, \tag{$\Psi$}$$

respectively.

We shall now define the indifference curves in the two regions. In the first region, the family of indifference curves has for its equation

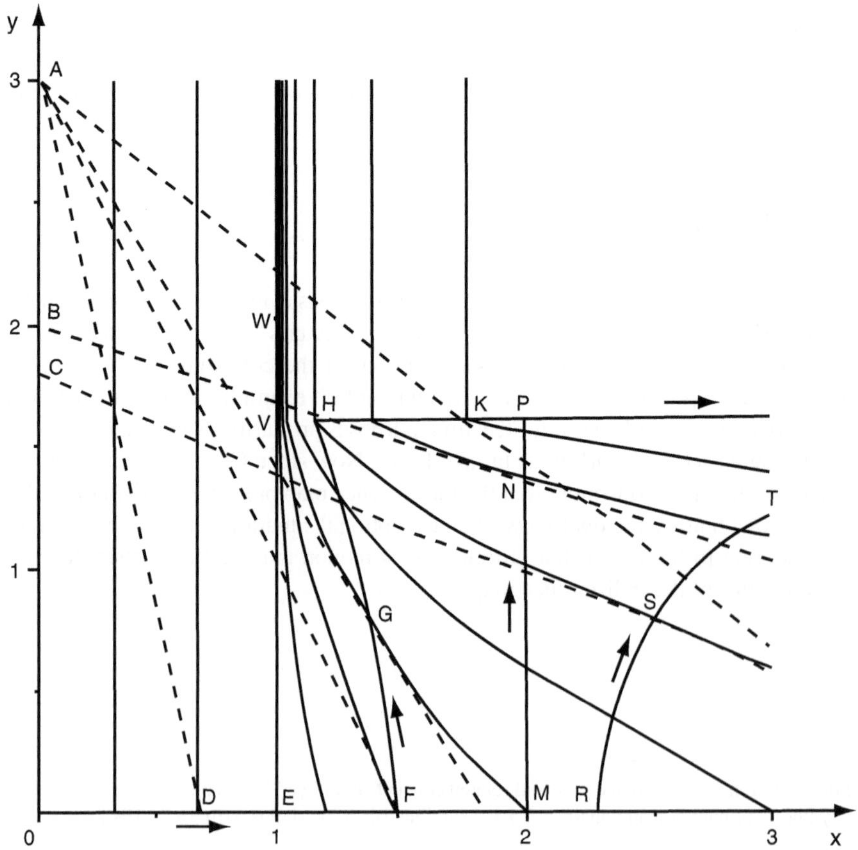


**Fig. 1** Indifference map

$$0 \cdot 16 \frac{x-1}{(y-2)^2} + 1 = c. \qquad (\Phi)$$

The graph shows 7 such curves, the $c$-values of which are $1 \cdot 008,\ 1 \cdot 02,\ 1 \cdot 04,\ 1 \cdot 08,\ 1 \cdot 16,\ 1 \cdot 4,\ 1 \cdot 8$; the last of these values has given the curve which passes through the point $K$.

In the second region, the indifference curves are defined by

$$x = c. \qquad (\Psi)$$

The indifference curves in this region thus are straight lines parallel to the $y$-axis.

On the border line VHKP ..., every indifference curve in the region $\Phi$ passes into a certain indifference curve in the region $\Psi$. Any two curves thus connected correspond with the same $c$-value in the equations ($\Phi$) and ($\Psi$), as is easily verified. The parameter $c$ may accordingly be interpreted as a utility index, $c = c(x, y)$, or a *preference index*, to use the term which I have suggested in order to avoid the notion of utility.[2]

*Remark 1.* The curve family defined by ($\Phi$) and ($\Psi$) fulfils the general requirements of an indifference map.[3] We note that in the region $x < 1$, where the indifference curves are parallel to the $y$-axis, the situation allows the interpretation that the consumer has no demand for the commodity $\eta$ if he can afford only a quantity $x < 1$ of the commodity $\xi$. In the region $x > 1, y > 1 \cdot 6$, where the curves are likewise parallel to the $y$-axis, the map may be interpreted according to any of the following alternatives:

(a) Provided that the consumer has a quantity $x \geq 1$ of the commodity $\xi$, his demand for $\eta$ is completely satisfied as soon as $y = 1 \cdot 6$;
(b) The commodity $\eta$ is rationed, with $y = 1 \cdot 6$ as maximal quantity for an individual consumer.

It is seen that in the region where $y < 1 \cdot 6$ our indifference map corresponds with the situation generally referred to when explaining Giffen's paradox: At a low standard of living, $\xi$ appears as a necessity, $\eta$ as a pure luxury; at higher income levels, however, the consumer's preference turns rapidly from $\xi$ to $\eta$.

## 3 Demand Functions: Determination

We are now in a position to determine the quantities $x = X, y = Y$ which the consumer will buy, given his income $\mu$ and the prices $p_x$ and $p_y$, say. The combination $X$, $Y$ is called the optimal budget. According to the classical theory

---

[2]Loc. cit., III, p. 74.

[3]Loc. cit., II, p. 221 ff.

of choice *X, Y* is a point on the budget line

$$xp_x + yp_y = \mu \tag{1}$$

and makes, subject to this condition, the preference index $c(x, y)$ of our indifference map a maximum. It is also well known that the optimal budget allows the representation

$$X = X\ (p_x/\mu,\ p_y/\mu),\quad Y = Y\ (p_x/\mu,\ p_y/\mu). \tag{2}$$

The functions $X$ and $Y$ being the demand functions for the two commodities, Giffen's phenomenon is present if one function, say $X$, decreases when $p_x$ is allowed to decrease, holding $\mu$ and $p_y$ constant.

Now to determine $X$ and $Y$ we make $\mu = 1$, which clearly involves no loss of generality, and allow $p_x$ to vary, while $p_y$ is held fixed. Since the budget line (1) intersects the $y$-axis at $y = 1/p_y$, the budget line will turn around this point as $p_x$ varies. For instance, *AD, AF, AG* and *AK* are budget lines with $p_y = {}^1/_3$.

The budget line (1) intersects the $x$-axis at $x = 1/p_x$. For $p_x \geq 1$ the intersection is in the interval (0,1). It is seen from the map that for any such line, say *AD*, the preference index will attain its maximum on the $x$-axis. Accordingly we obtain, for $p_x \geq 1$,

$$X = 1/p_x,\quad Y = 0. \tag{3}$$

In other words, as $p_x$ decreases from infinity down to $p_x = 1$, the optimal budget *X, Y* will describe the segment *ODE*.

As $p_x$ falls below 1, the optimal budget will still move rightwards along the $x$-axis, continuing until the budget line becomes a tangent of an indifference curve (for the budget lines through $A$, this occurs at $F$). Allowing $p_x$ to decrease further, the budget line will touch a certain indifference curve, and the tangent point will be the optimal budget (for instance, *G* is the optimal budget of the budget line *AG*). The tangent point is determined by the familiar system

$$\frac{\frac{\partial}{\partial x}c(x,\ y)}{p_x} = \frac{\frac{\partial}{\partial y}c(x,\ y)}{p_y};\quad xp_x + yp_y = 1. \tag{4}$$

Inserting the preference index $c(x, y)$ given by ($\Phi$), the system (4) gives after some easy calculation

$$X = \frac{2}{p_x}\left(p_x + p_y - \frac{1}{2}\right),\quad Y = \frac{2}{p_y}(1 - p_x - p_y). \tag{5}$$

Making here $Y = 0$, we obtain the price $p_x$ for which the demand functions shift from (3) into (5), viz.

$$p_x = 1 - p_y. \tag{6}$$

## 4 Demand Functions: Giffen's Paradox

The demand function $X$ as given by (5) provides an example of Giffen's paradox. In fact, we obtain

$$\frac{\partial X}{\partial p_x} = \frac{1 - 2\,p_y}{(p_x)^2},$$

which shows that $\partial X/\partial p_x \gtrless 0$ according as $p_y \lesseqgtr \frac{1}{2}$, and so $X$ will decrease with $p_x$ for any fixed $p_y < {}^1/_2$. Referring to the map for illustration, it is seen that if $p_y = {}^1/_3$, the demand functions shift from (3) to (5) for $p_x = {}^2/_3$; at this price the optimal budget is $F$, where $X = {}^3/_2$. Now if $p_x$ is allowed to decrease further, the optimal budget will describe the curve *FGH*. This implies that the quantity $X$ will here decrease with $p_x$, which constitutes Giffen's phenomenon.

As to the case when $p_y$ is fixed at a level $> \frac{1}{2}$, the graph shows a budget line *CS* for which $p_y = {}^5/_9$. The shift from (3) to (5) takes place at the point $R$, where $X = 2 \cdot 25$. As the fall of the price $p_x$ continues, the optimal budget will describe the curve *RST*. Thus $X$ continues to increase as $p_x$ decreases, and so Giffen's phenomenon is not present.

In the intermediary case $p_y = {}^1/_2$, the shifting is in the point $M$, where $p_x = {}^1/_2$ and $X = 2$; for further decreasing $p_x$ the optimal budget will move upwards along the vertical line *MNP*.

Returning to the case $p_y < {}^1/_2$, and paying regard to (6), we observe that as $p_x$ goes below $1 - p_y$ the optimal budget will follow a curve of the type *FG* up to the horizontal line $y = 1 \cdot 6$ and from there move to the right along the line *VHP*, the twist being of course due to the corners which the indifference curves present along the line *VHP* ... Thus Giffen's phenomenon is restricted to curve segments of the type *FGH*. The corresponding prices $p_x$ are those for which $Y$ in (5) varies between 0 and $1 \cdot 6$, i.e. the prices $p_x$ for which

$$1 - p_y \geq p_x \geq 1 - 1 \cdot 8\,p_y. \tag{7}$$

## 5 Giffen's Paradox and Offer Curves

Summing up, we have found that if the price $p_y$ is fixed below ${}^1/_2$, and $p_x$ falls from infinity down to zero, the optimal budget in our indifference map will describe a curve of the type *ODFGHKP* .... Giffen's paradoxon will be present in the segment *FGH*, i.e. for prices in the interval (7).

For prices $p_y \geq {}^1/_2$, our indifference map will not give rise to Giffen's phenomenon.

*Remark 2.* In the terminology of A.L. Bowley,[4] the curve described by the optimal budget as one price varies, say $p_x$, while the other price is held fixed – for instance,

[4] A.L. Bowley: The mathematical groundwork of economics. Oxford 1924.

curves such as *ODFGHKP* … and *ODRST* … — are called *offer curves*. The equation of the offer curves is obtained by eliminating $p_x/\mu$ from the demand relations (2). Thus if we eliminate $p_x$ from (5), we obtain the offer curves inside the region $\Phi$ (= the infinite rectangle … *RMFEVHKP* …), viz.

$$X = \frac{2(1 - Y\, p_y)}{2 - 2\, p_y - Y\, p_y}.$$

If we specify by inserting $p_y = {}^1/_3$, we obtain the equation of the offer curve *FGH*, viz.

$$X = \frac{6 - 2\,Y}{4 - Y}.$$

## 6 Revisiting the Indifference Map

The indifference curves in the region $\Phi$ of Fig. 1 are parabolae with $W$ for common apex. However, if the curves actually ran together in a certain point, the converging curves would clearly represent the same preference level, and so the curve family had to be discarded as an indifference map.[5] It is for this reason that we have cut off the region $\Phi$ by the line $y = 1 \cdot 6$. It is seen that this cut restricts the region where Giffen's phenomenon is present. Now such a limitation in the presence of the phenomenon is not confined to the particular map of Fig. 1. In fact, we have the following general

**Theorem 1.** *The offer curves which present Giffen's phenomenon cannot cover the whole of the indifference map.*

To establish this theorem we shall first prove the following

**Lemma 1.** *In a region where the offer curves form a system of straight lines parallel to the y-axis, the indifference map allows the representation*

$$y = f(x) + c.x. \tag{8}$$

Referring to Fig. 2a for illustration, let

$$y = f(x)$$

be the equation of an arbitrarily chosen indifference curve *rs*. The tangent at any point $(x,\ y)$ of this curve will cut the $y$-axis at the ordinate

$$f(x) - xf'(x), \tag{9}$$

as is easily verified.

[5]For further comments on this situation, see H. Wold, loc. cit., II, § 31, esp. p. 230.

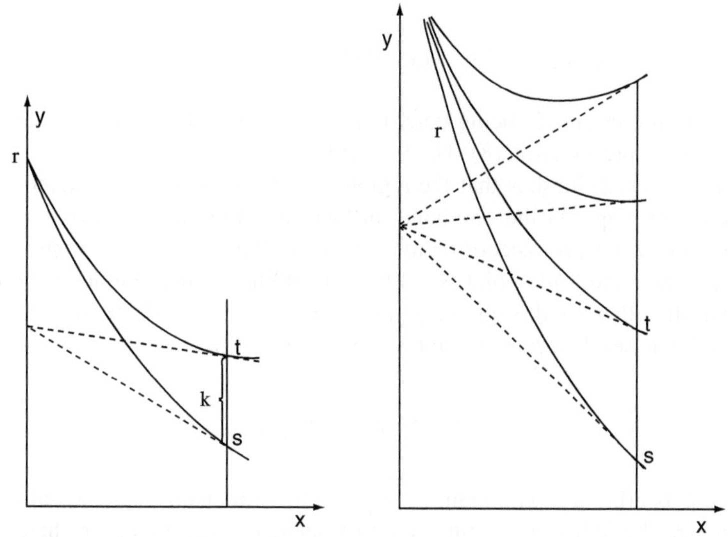


**Fig. 2** Illustration belonging to the proofs

Considering now the region of the lemma, let $t$ be a point on the offer curve which passes through $s$, let

$$y = \phi(x)$$

be the equation of the indifference curve which passes through $t$, and write

$$\phi(x_0) - f(x_0) = k \tag{10}$$

where $x_0$ is the common abscissa of $s$ and $t$. By the definition of offer curves, the tangents of the indifference curves at $s$ and $t$ will meet on the $y$-axis. Hence, paying regard to (9), we infer that the relation

$$f(x) - xf'(x) = \phi(x) - x\phi'(x) \tag{11}$$

holds good for $x = x_0$. More generally, since all offer curves are parallel to the $y$-axis, the relation (11) will be satisfied for all $x$.

The identity (11) thus established allows us to deduce $\phi$ $(x)$ from $f(x)$. In fact, the transformation $\phi = xz'$ carries (11) into

$$x^2 z'' = xf'(x) - f(x),$$

which gives

$$z' = \int \frac{xf'(x) - f(x)}{x^2} dx = \frac{f(x)}{x} + C.$$

Hence

$$\phi(x) = xz' = f(x) + Cx = f(x) + \frac{k}{x_0} \cdot x, \tag{12}$$

the integration constant $C$ being determined from (10). Now since $k$ is arbitrary, (12) gives the representation (8). The lemma is proved.

The relation (12) implies that the region of vertical offer curves cannot extend over the entire map. On the one hand, in fact, the indifference curve $y = f(x)$ cannot cut the $y$-axis, for according to (11) the indifference curve $y = \phi(x)$ would cut the $y$-axis in the same point (see Fig. 2a), and this is impossible, as remarked above. On the other hand, suppose $f(x) \to \infty$ as $x \to a$, where we can at once exclude other cases than $a = 0$ (see Fig. 2b). Now by (12),

$$\phi'(x) = f'(x) + \frac{k}{x_0},$$

and since $k$ is arbitrary, we obtain $\phi'(x) > 0$ for sufficiently large $k$-values. In a region where the indifference curves are of positive slope, however, there are no offer curves.[6]

Having now shown that an indifference map cannot be completely covered by vertical offer curves, our theorem follows as a corollary. In fact, it is easily seen that in a region where the offer curves present Giffen's paradox, the indifference curves will run together even more rapidly than in the case of vertical offer curves, and so the previous arguments will apply a fortiori.

[6]Cf. H. Wold, loc. cit., II, § 37, esp, p. 228.

# Giffen Demand for Several Goods

**Peter Norman Sørensen**

**Abstract** The utility maximizing consumer's demand function may simultaneously possess the Giffen property for any number of goods strictly less than all. By way of a simple example it is possible to illuminate the preference characteristics conducive to such a result.

## 1 Introduction

Recent examples of simple, standard utility functions can be used to obtain a fuller understanding of the Giffen effect in standard demand theory.[1] The Giffen effect arises when a consumer's demand for a good is locally increasing in the good's own price. Microeconomics textbooks often mention that the demand function for one good may possess the Giffen property; that Giffen goods must be inferior; and that not all goods can be simultaneously inferior. By implication, not all goods can be Giffen.[2]

Here, the converse statement is explored. Any number but one of the goods can simultaneously have the Giffen property. A particular price-income pair is held fixed, and the demand for each good in question is locally a strictly increasing function of its own price.[3]

[1]Heijman and Mouche [1] surveys the literature.

[2]For completeness, a new version of this result is spelled out towards the end of this article.

[3]Holding fixed only the price vector, but allowing the income to change, different goods can have the Giffen property at different income levels. In this weaker sense it is possible for all goods to exhibit the Giffen property at some price vector. An explicit example appears in Example 5 of [7].

P.N. Sørensen (✉)
University of Copenhagen, Øster Farimagsgade 5 Building 26, 1353 Copenhagen K, Denmark
e-mail: peter.sorensen@econ.ku.dk

W. Heijman and P. Mouche (eds.), *New Insights into the Theory of Giffen Goods*, Lecture Notes in Economics and Mathematical Systems 655,
DOI 10.1007/978-3-642-21777-7_8, © Springer-Verlag Berlin Heidelberg 2012

Generalizing the main idea from [8]'s two-good analysis, the present article considers a consumer with utility function $u(x) = \min\{u_1(x), \ldots, u_L(x)\}$ where $u_1, \ldots, u_L$ are themselves standard utility functions over $L$-good consumption bundles $x = (x_1, \ldots, x_L)$. The utility function $u$ represents preferences for perfect complements in the intermediate utility indices $u_1, \ldots, u_L$. Bearing in mind that the most popular examples of utility functions (perfect substitutes, perfect complements, Cobb–Douglas, constant elasticity of substitution) derive the utility index as some weighted average of the consumed amounts $x_1, \ldots, x_L$, it is natural to compound the average of such averages.[4]

Along the lines of [8], the focus is on the budget situations where the consumer optimally demands a bundle $\hat{x}$ at an intersection of the $L$ indifference surfaces, $u_1(\hat{x}) = \cdots = u_L(\hat{x}) = \hat{u}$. At such a corner of the indifference surface $u(x) = \hat{u}$, there is no substitution effect as the price changes. The consumer's reaction to a small price increase is then equivalent to the reaction to a small income reduction.[5] A good now has the local Giffen property if and only if it is locally inferior. The only remaining problem is then to place a backward-bending income expansion path, which is analytically simple.[6]

## 2 Utility

Let $J, L \in \mathbb{N}$. Consider a consumer with consumption set $\mathbb{R}_+^L$ whose utility function satisfies $u(x) = \min\{u_1(x), \ldots, u_J(x)\}$, where each $u_j$ is a function from $\mathbb{R}_+^L$ to $\mathbb{R}$. The utility function $u$ from $\mathbb{R}_+^L$ to $\mathbb{R}$ satisfies many usual properties if each of the component utility functions $u_j$ does so. Here is a list of such properties (see also Section 3.B of [4]). Property (vi) is also known as convexity of the underlying preference relation.

(i) Continuity: $u$ is a continuous function.
(ii) Monotonicity: if $y \gg x$, then $u(y) > u(x)$.
(iii) Strong Monotonicity: if $y \geq x$ and $y \neq x$ then $u(y) > u(x)$.
(iv) Weak Monotonicity: if $y \geq x$ then $u(y) \geq u(x)$.
(v) Concavity: $u$ is a concave function.
(vi) Quasi-Concavity: for any $\bar{u} \in \mathbb{R}$, the upper contour set $\{x \in \mathbb{R}_+^L : u(x) \geq \bar{u}\}$ is convex.

[4] This construction of utility functions has been suggested before, see e.g. exercise 8.13 of [9] and exercises 1.12(b) and 1.27 of [2].

[5] Figure 17.E.1 of [4] illustrates preferences of a very similar nature. The expansion curves are parallel to the axes, providing a Giffen neutrality effect, where the demand for a good does not respond to a marginal change in its own price.

[6] Constructing a utility function with Giffen demand, [5] proceeds in the reverse direction. He first takes the backward-bending income expansion path, and then attaches a string of nearly-kinked indifference curves to it.

(vii) Strict Quasi-Concavity: for any $\bar{u} \in \mathbb{R}$, any $x, y \in \mathbb{R}_+^L$ with $x \neq y$, and any $\alpha \in (0,1)$, if $u(x) \geq \bar{u}$ and $u(y) \geq \bar{u}$, then $u(\alpha x + (1-\alpha) y) > \bar{u}$.

(viii) Homogeneity of Degree 1: $u$ maps $\mathbb{R}_+^L$ into $\mathbb{R}_+$, and $u(\alpha x) = \alpha u(x)$ for all $\alpha \in \mathbb{R}_+$.

It is easy to verify that each of those eight properties, one by one, is inherited by $u$ from $u_1, \ldots, u_J$:

**Lemma 1.** *Let $n \in \{i, \ldots, viii\}$. Suppose that $u_1, \ldots, u_J$ all satisfy property $(n)$. Then $u$ satisfies $(n)$.*

*Proof.* (i) min is a continuous function, and continuity is preserved by the composition of functions. (ii) and (iii) Suppose $y$ and $x$ are as assumed in the property. By definition of $u$ there exists some $j$ such that $u(y) = u_j(y)$. Given that $u_j$ satisfies the property, then $u(x) \leq u_j(x) < u_j(y) = u(y)$. (iv) Similar to (ii) and (iii), except that the assumption is $u_j(x) \leq u_j(y)$, which suffices for the conclusion. (v) Let $x, y \in \mathbb{R}_+^L$ and $\alpha \in [0,1]$, and verify Jensen's inequality: $u(\alpha x + (1-\alpha) y) = \min\{u_1(\alpha x + (1-\alpha) y), \ldots, u_J(\alpha x + (1-\alpha) y)\} \geq \min\{\alpha u_1(x) + (1-\alpha) u_1(y), \ldots, \alpha u_J(x) + (1-\alpha) u_J(y)\} \geq \alpha u(x) + (1-\alpha) u(y)$, where the first inequality uses the concavity of the $u_j$ functions with min being increasing in its arguments, and the second uses concavity of min. (vi) Suppose that $x \in \mathbb{R}_+^L$ and $\bar{u} \in \mathbb{R}$. From the definition of $u$, $u(x) \geq \bar{u}$ if and only if for every $j$, $u_j(x) \geq \bar{u}$. Thus the upper contour set for $u$ is the intersection of the $J$ sets for $u_1, \ldots, u_J$. Since the intersection of convex sets is convex, (vi) follows. (vii) Suppose that $\bar{u}, x, y, \alpha$ are given as stated, and that $u(x), u(y) \geq \bar{u}$. By the definition of $u$, there exists some $j$ such that $u(\alpha x + (1-\alpha) y) = u_j(\alpha x + (1-\alpha) y)$. As noticed in the proof of (vi), we must have $u_j(x), u_j(y) \geq \bar{u}$. Since $u_j$ satisfies (vii), it follows that $u(\alpha x + (1-\alpha) y) = u_j(\alpha x + (1-\alpha) y) > \bar{u}$ as desired. (viii) We obtain $u(\alpha x) = \min\{u_1(\alpha x), \ldots, u_J(\alpha x)\} = \min\{\alpha u_1(x), \ldots, \alpha u_J(x)\} = \alpha \min\{u_1(x), \ldots, u_J(x)\} = \alpha u(x)$. □

If the consumption set allows good 1 to vary throughout all of $\mathbb{R}$, a straightforward exercise proves that Lemma 1 also applies to this property of quasi-linearity with respect to good 1: $u(x + \alpha e_1) = u(x) + \alpha$ for all $\alpha \in \mathbb{R}$, where $e_1 = (1, 0, \ldots, 0)$.

Lemma 1 can be generalized to handle the minimum of an infinite family of utility functions. Suppose thus that $U(x) = \min_{j \in I} u(x, j)$ where $I$ is a compact set and $u$ is continuous. These assumptions guarantee that the minimum is achieved, but also imply that $U$ is continuous by the Theorem of the Maximum. It is a straightforward exercise to see that the proof of Lemma 1 can be modified to show that the other seven properties (ii),…,(viii) of $u(\cdot, j)$ are again inherited by $U$.

The following standard result will be useful in the remainder.

**Lemma 2.** *Suppose that $u$ is a function from $\mathbb{R}_+^L$ to $\mathbb{R}$ satisfying (i) continuity, (ii) monotonicity and (vi) quasi-concavity. For any $x \in \mathbb{R}_+^L$, there exists $p \in \mathbb{R}_+^L$ such* *that $u(y) > u(x)$ implies $p \cdot y > p \cdot x$. If $u$ is differentiable at $x$ with $\nabla u(x) \neq 0$, then $p$ must be proportional to $\nabla u(x)$.*

*Proof.* First, let $A = \{y \in \mathbb{R}^L_+ | u(y) > u(x)\}$. Properties (i) and (vi) imply that $A$ is open and convex, and $x \notin A$ by construction. A standard separation theorem for convex sets, such as 11.2 in [6], yields the existence of $p \in \mathbb{R}^L$ with $p \cdot y > p \cdot x$ for all $y \in A$. Also, $p \geq 0$ is necessary since (ii) implies that $\{x\} + \mathbb{R}^L_{++} \subseteq A$. Second, suppose that $u$ is differentiable at $x$ with $\nabla u(x) \neq 0$. If $p$ is not proportional to $\nabla u(x)$, then there exists $z \in \mathbb{R}^L$ with $p \cdot z < 0 < \nabla u(x) \cdot z$. By differentiability, there exists $\varepsilon > 0$ with $u(x + \varepsilon z) > u(x)$, so that $x + \varepsilon z \in A$. The separation result then implies $p \cdot (x + \varepsilon z) > p \cdot x$, in contradiction to $p \cdot z < 0$. □

## 3 Demand

Suppose $L \geq 2$. Given income $m > 0$ and price vector $p = (p_1, \ldots, p_L) \in \mathbb{R}^L_{++}$, the consumer chooses $x = (x_1, \ldots, x_L)$ to maximize utility $u(x)$ subject to the budget constraint $p \cdot x = p_1 x_1 + \cdots + p_L x_L \leq m$. The maximand is the Marshallian demand $x(p, m)$.

The Marshallian demand for good $\ell$ possesses the Giffen property at price-income pair $(p, m)$ if there is a strictly positive marginal response in $x_\ell(p, m)$ to a partial change in $p_\ell$.

In the following, we will suppose that $J = L$ and that the consumer's utility function satisfies $u(x) = \min\{u_1(x), \ldots, u_L(x)\}$. We wish to study the consumer's demand function near a situation where the demanded bundle $x$ sits at a kink of the indifference surface for $u$. Naturally, such a kink arises at $x$ where $u_1(x) = \cdots = u_J(x)$ and the gradient vectors $\nabla u_1(x), \ldots, \nabla u_J(x)$ are linearly independent. Note that such a point $x$ would not define a kink of the indifference surface if $J < L$, and that $J - L$ of the functions $u_j$ would be redundant in the definition of the utility function $u$ near $x$ if $J > L$. Hence, the remainder of the text assumes $J = L$.

**Proposition 1.** *Suppose that $u_1, \ldots u_L$ all satisfy (i) continuity, (ii) monotonicity and (vi) quasi-concavity. Assume that $\hat{x} \in \mathbb{R}^L_{++}$ solves $u_1(\hat{x}) = \cdots = u_L(\hat{x})$, that $u_1, \ldots, u_L$ are $C^1$ at $\hat{x}$, and that the gradient vectors $\nabla u_1(\hat{x}), \ldots, \nabla u_L(\hat{x})$ are linearly independent. Take as given any price vector $\hat{p} = \lambda_1 \nabla u_1(\hat{x}) + \cdots + \lambda_L \nabla u_L(\hat{x})$ with $(\lambda_1, \ldots, \lambda_L) \gg 0$, and let $\hat{m} = \hat{p} \cdot \hat{x}$. Define $\hat{d} \in \mathbb{R}^L \setminus \{0\}$ by*

$$\hat{d} = \left[1 \cdots 1\right] \left[\nabla u_1(\hat{x}) \mid \cdots \mid \nabla u_L(\hat{x})\right]^{-1}. \tag{1}$$

*If $\hat{d}_\ell < 0$, then good $\ell$ is a Giffen good near $(\hat{p}, \hat{m})$ for the consumer with utility function $u$.*

*Proof.* First, observe that $\hat{x}$ is a solution to the utility maximization problem given $(\hat{p}, \hat{m})$. Define $\hat{u} \in \mathbb{R}$ by $\hat{u} = u_1(\hat{x})$. By definitions, $\hat{x}$ is feasible in the problem and achieves utility level $\hat{u}$. Consider any $y \in \mathbb{R}^L_+$ with $u(y) > \hat{u}$. Observe that for any $j$, $u_j(y) \geq u(y) > \hat{u} = u_j(\hat{x})$. By Lemma 2 applied to $u_j$, then $\nabla u_j(\hat{x}) \cdot y > \nabla u_j(\hat{x}) \cdot \hat{x}$. Since every $\lambda_j > 0$, it follows that $\hat{p} \cdot y > \hat{p} \cdot \hat{x} = \hat{m}$, so any such $y$ lies outside the budget set. Second, by the Theorem of the Maximum, the maximal utility that the consumer can achieve is a continuous function of $(p, m)$. Since the utility functions $u_j$ are $C^1$, and since the gradient vectors are linearly independent, the first part of the proof extends to an open region of price-income pairs $(p, m)$ near $(\hat{p}, \hat{m})$, in which region a solution $x(p, m)$ near $\hat{x}$ to the utility maximization problem solves the equations $u_1(x(p, m)) = \cdots = u_L(x(p, m)) = \tilde{u}(p, m)$ for some $\tilde{u}(p, m)$ near $\hat{u}$. Third, an increase in the price of good $\ell$ will ceteris paribus decrease the utility level, so the demanded bundle $x$ moves along the curve solving $u_1(x) = \cdots = u_L(x) = \tilde{u}$ for decreasing utility level $\tilde{u}$. The claim is then that $x_\ell$ is a decreasing function of $\tilde{u}$ when $\hat{d}_\ell < 0$. This follows from an application of the implicit function theorem to the $L$ equations $u_1(x) = \tilde{u}, \ldots, u_L(x) = \tilde{u}$. □

The proposition actually verifies a stronger property than Giffen's when more than one good is involved. If the entries in $\hat{d}$ are negative for all of some subset $K$ of the goods, then holding fixed the income as well as prices of all other goods, the Marshallian demand for all $K$ goods in the subset will increase when all their prices are simultaneously marginally increased.

The proposition exploits the zero substitution effect at a kink in the indifference curve. This extreme can be relaxed through approximation. The function $\min\{u_1, \ldots, u_L\}$ is approximated by the constant elasticity of substitution (CES) function $\left(u_1^\rho + \cdots + u_L^\rho\right)^{1/\rho}$ as $\rho \to -\infty$. If $u_1, \ldots, u_L$ are concave functions of $x$, then also $u(x) = \left(u_1^\rho(x) + \cdots + u_L^\rho(x)\right)^{1/\rho}$ is concave in $x$. The function $u$ therefore represents continous, monotone, convex preferences. When $-\rho$ is sufficiently large, all indifference curves near $\hat{x}$ are sufficiently close to those of $\min\{u_1(x), \ldots, u_L(x)\}$, and the sign of $\hat{d}$ determines whether the good is Giffen also in the CES case.

## 4 Example

An example will illustrate how the proposition can be applied to construct demand functions with local Giffen behavior. Let $L \geq 2$. Suppose that each $u_j$ is of the familiar Cobb–Douglas form

$$u_j(x_1, \ldots, x_L) = x_1^{a_{j1}} \cdots x_L^{a_{jL}}, \tag{2}$$

with strictly positive constants $a_{j1}, \ldots, a_{jL}$. This utility function satisfies (i) continuity, (ii) monotonicity and (vi) quasi-concavity. Specifically, we will suppose that the vectors $a_1, \ldots, a_L \in \mathbb{R}^L_+$ are defined by $a_j = (1, \ldots, 1, b, 1, \ldots, 1)/c_j$ with $b > 1$ in the $j$'th coordinate, and $c_j \geq b + L - 1$. Since $a_{j1} + \cdots + a_{jL} \leq 1$, the utility function $u_j$ is concave.

For the sake of the example, focus will be on the consumption bundle $\hat{x} = (1, \ldots, 1)$. For $\hat{u} = 1$, we have $u_j(\hat{x}) = \hat{u}$, and $u_j$ is $C^1$ at $\hat{x}$ with $\nabla u_j(\hat{x}) = a'_j$. The assumptions of the proposition are satisfied, so calculation of the vector $\hat{d}$ helps to determine which are Giffen goods at $\hat{x}$. It is simple to verify that

$$\left[\nabla u_1(\hat{x}) \mid \cdots \mid \nabla u_L(\hat{x})\right]^{-1} = G \tag{3}$$

where row $j$ of matrix $G$ is

$$g_j = \frac{c_j}{b(b+L-2)-(L-1)}[-1, \ldots, -1, b+L-2, -1, \ldots, -1] \tag{4}$$

with $b+L-2$ in the $j$'th coordinate. Note that $b > 1$ and $L \geq 2$ imply $b+L-2 > 1$ as well as $b(b+L-2)-(L-1) > 0$.

Now, it follows from (1) that

$$\hat{d}_\ell = \sum_{j=1}^{L} g_{j\ell} = \frac{c_\ell(b+L-2) - \sum_{j\neq\ell} c_j}{b(b+L-2)-(L-1)}. \tag{5}$$

Since the denominator is positive, it should be clear that the sign of $\hat{d}_\ell$ is determined by the relative size of $c_\ell$ to $\sum_{j\neq\ell} c_j$. In particular, $\hat{d}_\ell < 0$ and $\ell$ is a local Giffen good around the price-income pairs defined in the proposition, when $c_\ell$ is sufficiently relatively small. The only assumption is that each such positive constant is at least $b+L-1$, so it is not hard to make all but one such constant relatively small (by making the last one very large).

For an illustration, let $1 \leq k < L$ and suppose that $b+L-1 \leq c_1 = \ldots = c_k < c_{k+1} = \ldots = c_L$. Each of the first $k$ goods will simultaneously be local Giffen goods provided $c_1(b+L-2-k+1) < (L-k)c_L$. This is true when $c_L$ is sufficiently large.

A natural interpretation of the utility function $\min\{u_1(x), \ldots, u_L(x)\}$ follows from [3]'s new consumer theory, in which consumers care about characteristics produced by the purchased bundle. In the present context, think of the concave Cobb–Douglas functions as production functions.[7] The assumption that $b > 1$ means that factor $\ell$ is relatively essential for the production of characteristic $\ell$. The illustration then provides the important insight that, given the existence of a kink point $\hat{x}$, good $\ell$ is more likely to be Giffen if $c_\ell$ is relatively large. This means that the production of characteristic $\ell$ offers a relatively high marginal return to all its inputs. The following intuitive reason can then explain the Giffen property. When the price of good $\ell$ rises, the production of all characteristics will be scaled back. Since characteristics are perfect complements, those with lower marginal returns

[7] This interpretation does not permit the consumer to allocate inputs in different proportions across productive activities. Lancaster [3] formulates the model both with and without such a possibility.

can scale back their relatively essential factors a lot for a small reduction in the characteristic. In turn, this scaling back would give too great a reduction in the high-marginal return characteristic $\ell$ if it were not for an increase in its relatively essential factor $\ell$.

## 5 Converse

As already mentioned in the introduction, it is commonly accepted that there cannot exist a price-income pair at which all $L$ goods are simultaneously Giffen. The standard derivation of this result goes via the Slutsky equation. The argument thus invokes an assumption of differentiability of the demand function which needs not always be satisfied. For the sake of completeness, here is a result derived from first principles.

The solution to the utility maximization problem must satisfy the following weak axiom of revealed preferences (WARP). Given two pairs $(p, m), (p', m') \in \mathbb{R}^{L+1}_{++}$, if $x(p, m) \neq x(p', m')$ and if $p \cdot x(p', m') \leq m$ then $p' \cdot x(p, m) > m'$. Interpreting this, $p \cdot x(p', m') \leq m$ implies that $x(p', m')$ is a feasible choice at $(p, m)$, so that utility maximization reveals the preference relation $u(x(p, m)) > u(x(p', m'))$.[8]

For convenience, it will be assumed that the demand function is continuous. The only purpose of this assumption in the argument is to ensure that there is a neighborhood around the hypothetical $L$ Giffen good point $(p, m)$ in which the demand function is bounded away from zero in all coordinates. Continuity of the demand function can be derived as a necessary implication of properties (i) continuity and (vii) strict quasi-concavity of the utility function.

**Proposition 2.** *Suppose that the demand function is continuous. There does not exist any* $(p, m) \in \mathbb{R}^{L+1}_{++}$ *with the property that for all* $\ell \in \{1, \ldots, L\}$, *good* $\ell$ *is Giffen in a neighborhood of* $(p, m)$.

*Proof.* Suppose to the contrary that such $(p, m) \in \mathbb{R}^{L+1}_{++}$ exists. The proof will seek a contradiction. Denote $x^0 = x(p, m)$. Observe that $x^0 \gg 0$ since the non-negative consumption of every good will be strictly smaller when its own price falls. Observe also that $p \cdot x^0 = m$, for there would otherwise exist $\hat{p} = p + \varepsilon p_1$ close enough to $p$ with $\varepsilon > 0$ such that $\hat{p} \cdot x^0 \leq m$, in contradiction to the WARP because $x(\hat{p}, m) \neq x^0$ and $p \cdot x(\hat{p}, m) \leq \hat{p} \cdot x(\hat{p}, m) \leq m$. Denote $\ell^1 = 1$. Choose $p^1 \in \mathbb{R}^L_{++}$ close to $p$ such that $p^1_\ell = p_\ell$ for all $\ell \neq \ell^1$, and $p^1_{\ell^1} > p_{\ell^1}$. Denote $x^1 = x(p^1, m)$. Observe that $x^1_{\ell^1} > x^0_{\ell^1} \geq 0$ since $\ell^1$ is a local Giffen good. Also, $p \cdot x^1 < p^1 \cdot x^1 \leq m = p \cdot x^0$, ruling out $x^1 \geq x^0$. Hence, there exists some $\ell^2 \in \{2, \ldots, L\}$ with the property that $x^1_{\ell^2} < x^0_{\ell^2}$. Next, choose $p^2 \in \mathbb{R}^L_{++}$ such that $p^2_\ell = p_\ell$ for all $\ell \neq \ell^2$, and $\left(p^1_{\ell^1} - p_{\ell^1}\right) x^1_{\ell^1} =$

[8]See for instance Chaps. 1–3 in [4] for details.

$\left(p^2_{\ell 2} - p_{\ell^2}\right) x^1_{\ell^2}$. Denote $x^2 = x\left(p^2, m\right)$. Note that $p^2$ is arbitrarily close to $p$ when $p^1$ was chosen close enough to $p$. By construction, $p^2 \cdot x^1 = p_{\ell^1} x^1_{\ell^1} + p^2_{\ell^2} x^1_{\ell^2} + \sum_{\ell \neq \ell^1, \ell^2} p_\ell x^1_\ell = p^1_{\ell^1} x^1_{\ell^1} + p_{\ell^2} x^1_{\ell^2} + \sum_{\ell \neq \ell^1, \ell^1} p_\ell x^1_\ell = p^1 \cdot x^1 \leq m$. Since $\ell^2$ is a local Giffen good, $x^2_{\ell^2} > x^0_{\ell^2} > x^1_{\ell^2}$, so $x^2 \neq x^1$ . By the WARP, we have $u\left(x^2\right) > u\left(x^1\right)$ and $p^1 \cdot x^2 > m \geq p^2 \cdot x^2$. Expanding this inequality, observe that $\left(p^1_{\ell^1} - p_{\ell^1}\right) x^2_{\ell^1} > \left(p^2_{\ell^2} - p_{\ell^2}\right) x^2_{\ell^2} > \left(p^2_{\ell^2} - p_{\ell^2}\right) x^1_{\ell^2} = \left(p^1_{\ell^1} - p_{\ell^1}\right) x^1_{\ell^1}$, implying $x^2_{\ell^1} > x^1_{\ell^1} > x^0_{\ell^1}$. Re-iterating the previous argument, $x^2 \geq x^0$ is impossible, so there must exist some $\ell^3 \neq \ell^1, \ell^2$ with $x^2_{\ell^3} < x^0_{\ell^3}$. Repeat this step inductively, until a final $L$th iteration, in which $p^L \in \mathbb{R}^L_{++}$ is defined by $p^L_\ell = p_\ell$ for all $\ell \neq \ell^{L-1}$, and $\left(p^{L-1}_{\ell^{L-1}} - p_{\ell^{L-1}}\right) x^{L-1}_{\ell^{L-1}} = \left(p^L_{\ell^L} - p_{\ell^L}\right) x^{L-1}_{\ell^L}$. Continuity of the demand function implies that if the first choice of $p^1$ is sufficiently close to $p$, then every $p^2, \ldots, p^L$ constructed in this fashion is so close to $p$ that all the Giffen properties hold. Again, $u\left(x^L\right) > u\left(x^{L-1}\right)$ by the WARP, implying $x^L_{\ell^{L-1}} > x^{L-1}_{\ell^{L-1}} > x^0_{\ell^{L-1}}$. The Giffen good property of $L$ and the result of the previous induction step give $x^L_{\ell^L} > x^0_{\ell^L} > x^{L-1}_{\ell^L}$. Also, for every $z \in \{1, \ldots, L-1\}$, $u\left(x^L\right) > u\left(x^{L-z}\right)$ implies $p^{L-z} \cdot x^L > m = p^L \cdot x^L$. This is expanded as $\left(p^{L-z}_{\ell^{L-z}} - p_{\ell^{L-z}}\right) x^L_{\ell^{L-z}} > \left(p^L_{\ell^L} - p_{\ell^L}\right) x^L_{\ell^L} > \left(p^L_{\ell^L} - p_{\ell^L}\right) x^{L-1}_{\ell^L} = \left(p^{L-1}_{\ell^{L-1}} - p_{\ell^{L-1}}\right) x^{L-1}_{\ell^{L-1}} > \cdots > \left(p^{L-z}_{\ell^{L-z}} - p_{\ell^{L-z}}\right) x^{L-z}_{\ell^{L-z}}$. Hence, $x^L_{\ell^{L-z}} > x^{L-z}_{\ell^{L-z}} > x^0_{\ell^{L-z}}$ for every $z \in \{1, \ldots, L-1\}$. Since also $x^L_{\ell^L} > x^0_{\ell^L}$, it follows that $x^L \gg x^0$. This is the desired contradiction to $p \cdot x^L < p^L \cdot x^L = m = p \cdot x^0$. □

## References


1. Heijman, W and Mouche, PHM von (2011) A Child Garden of Concrete Giffen Utility Functions: a Theoretical Review. 655:65–81.
2. Jehle GA, Philip JR (2001) Advanced Microeconomic Theory, Second Edition. Addison-Wesley, London.
3. Lancaster KJ (1966) A New Approach to Consumer Theory, J Polit Econ 74:132–157.
4. Mas-Colell A, Whinston MD, Green JR (1995) Microeconomic Theory. Oxford University Press, Oxford.
5. Moffatt PG (2002) Is Giffen behaviour compatible with the axioms of consumer theory, J Math Econ 37:259–267.
6. Rockafellar RT (1970) Convex Analysis. Princeton University Press, Princeton.
7. Sørensen PN (2004) Simple Utility Functions with Giffen Demand, discussion paper 04-22. University of Copenhagen.
8. Sørensen PN (2007) Simple Utility Functions with Giffen Demand, Econ Theory 31:367–370.
9. Varian HR (1992) Microeconomic Analysis, Third Edition. Norton, New York.

# Giffen Behavior Independent of the Wealth Level

**Junko Doi, Kazumichi Iwasa, and Koji Shimomura**

Comment from the editors: this article concerns a reprint from *Economic Theory*, Volume 41(2), pages 247–267, 2009.

**Abstract** We demonstrate that a well-behaved utility function can generate Giffen behavior, where "well-behaved" means that its indifference curves are smooth, convex, and closed in a commodity space; the resulting demand function of each good is differentiable with respect to prices and income. Moreover, we show that Giffen behavior is compatible with any level of utility and an arbitrarily low share of income spent on the inferior good. This contrasts sharply with the common view that the Giffen paradox tends to occur when households' wealth levels are low.

## 1 Introduction

The Giffen paradox is usually construed as a possible economic behavior exhibited by households with low real wealth levels. As a recent example, a standard textbook of microeconomic theory provides an intuitive explanation of Giffen behavior as follows:

J. Doi (✉)
Faculty of Economics, Kansai University, 3-3-35 Yamate-cho Suita, Osaka, Japan
e-mail: jkdoi@ipcku.kansai-u.ac.jp

K. Iwasa
KIER, Kyoto University, Yoshida-Honmachi, Sakyo-Ku, Kyoto 606-8501, Japan
e-mail: kazumichi@hi-net.zaq.ne.jp

K. Shimomura
(† 2007) Kobe University, Japan.

W. Heijman and P. Mouche (eds.), *New Insights into the Theory of Giffen Goods*, Lecture Notes in Economics and Mathematical Systems 655,
DOI 10.1007/978-3-642-21777-7_9, © Springer-Verlag Berlin Heidelberg 2012

> Low-quality goods may well be Giffen goods for consumers with low wealth levels. For example, imagine that a poor consumer initially is fulfilling much of his dietary requirements with potatoes because they are a low-cost way to avoid hunger. If the price of potatoes falls, he can then afford to buy other, more desirable foods that also keep him from being hungry. His consumption of potatoes may well fall as a result. Note that the mechanism that leads to potatoes being a Giffen good in this story involves a wealth consideration: When the price of potatoes falls, the consumer is effectively wealthier (he can afford to purchase more generally), and so he buys fewer potatoes ... [3, p. 26].

This explanation of the Giffen paradox stems from the usual interpretation of the Slutsky equation: this equation shows that the Giffen effect arises only if the income effect is negative and exceeds the pure substitution effect in absolute value; such a large negative income effect is apt to arise when a household devotes a large share of an extremely low real income to one inferior good. Indeed, this equation has made some empirical researchers search for the Giffen paradox among goods consumed in large amounts.

The purpose of this paper is to provide a theoretical example that clearly shows the existence of Giffen behavior that is far removed from the intuition that the behavior is attributable to extremely low wealth levels and/or a high expenditure share on an inferior good. Specifically, we demonstrate a smooth, non-decreasing, quasi-concave utility function that has the following properties: First, the utility maximization problem has a unique, positive, and interior optimal consumption bundle; second, there exists a nontrivial subset of price–income pairs where one good (good 1) is a Giffen good; third, its Giffen behavior is possible, *irrespective of the level of utility; and finally, the behavior is compatible with an arbitrarily low share of income spent on that good, and not with a high share of income spent on that good.* Thus, Giffen behavior under this utility function reveals the theoretical possibility that it is inappropriate to presume the Giffen paradox to be merely an exceptional phenomenon that occurs only under extreme circumstances.

Let us briefly review the theoretical literature on the Giffen paradox. According to Weber [11], Wold and Jureen [12] were the first to present a specific utility function that generates Giffen behavior. Vandermeulen [9] and Spiegel [8] presented other specific utility functions with the Giffen property. A common feature of all of these proposed utility functions is that they need to be defined piecewise, that is, spliced with two or more functional forms to satisfy non-decreasingness or quasi-concavity.

Although these functions are not strictly quasi-concave over the whole commodity space, Moffatt [4] shows that for an arbitrary point on a two-dimensional (2D) commodity space, there exists a strictly quasi-concave utility function that generates a backward sloping price offer curve around it. Most recently, Sørensen [7] demonstrates a simple way of constructing a utility function with the Giffen property and provides an example of the utility function that is strictly quasi-concave and generates Giffen behavior. He splices two functional forms such that each indifference curve has a kink, which easily generates the Giffen effect because the substitution effect is equal to zero at that point.[1]

The purpose of the early studies was to show simple examples or the existence of utility functions with the Giffen property[2]; however, the utility functions they proposed are not suitable for a global analysis of demand for a Giffen good. Indeed, there has been no global analysis of the demand for a Giffen good. However, a specific function we will present makes this global analysis possible. The advantage of the function is as follows: although this function, too, is defined piecewise, unlike the specific functions proposed in the previous studies, we can consider only one functional form when the utility maximization problem is solved and obtain an interior solution where the first-order conditions are met with equalities; this makes the analysis of demand tractable.[3]

In contrast with the early papers, we derive the demand function of each good directly and thoroughly investigate it, both analytically and graphically. First, we characterize the domain of the demand functions and derive three subsets of the price–income pairs and the corresponding three subsets of consumption bundles as follows: in the first one, good 1 is a Giffen good; in the second, it is an inferior good, not a Giffen good; and in the third, it is a normal good. Based on these results, for an arbitrarily given point on the 2D positive commodity space, we can draw the income expansion path and the price offer curve (see Fig. 3B in this paper), both of which intersect with each other at the point. Moreover, however high households' income or utility level becomes along each income expansion path, we show that good 1 remains a Giffen good despite the fact that the households' expenditure share on the good is approaching zero. Thus, we can examine asymptotic behavior of demand for a Giffen good, which is impossible without specifying utility functions that generate Giffen behavior: There necessarily exists some limitation in drawing an indifference map that shows the Giffen paradox.

---

[1]Considering a general equilibrium model, Nachbar [5] points out that there is a fundamental difficulty in conducting empirical analyses on Giffen goods, since price changes in important goods often result in changes in households' income. However, in the case of Sørensen's example, the Giffen effect arises at the point where the substitution effect is zero, and hence, the amount of goods consumed is irrelevant.

[2]Using a specific utility function, Vandermeulen [9] also shows that expenditure shares are irrelevant to Giffen behavior because Giffen behavior depends solely on a property of the indifference map, i.e., the slope of indifference curves along vertical lines, and conjectures that a good is predisposed toward a Giffen good when it is a necessity and a household is rapidly satiated with its consumption.

[3]This yields another difference between the function we will propose and the other functions: the demand curve based on the former is smooth over the whole domain, while those based on the latter have a kink, where their slope changes from negative to positive discontinuously when the price falls below its threshold value. One can confirm that this discontinuity occurs because when one good turns into a Giffen good, the optimal consumption bundle changes from a corner solution to an interior one, or from an interior one where the first-order conditions are met with equalities to another interior one where they are not.

This paper is organized as follows. Section 2 presents the specific utility function. Section 3 proves that the utility maximization problem has a unique, positive, and interior optimal consumption bundle, and derives the demand function of each good. Section 4 characterizes the demand functions and proves the demand properties discussed above. Section 5 discusses some possible generalizations of the utility function. Section 6 provides the conclusion.

# 2 Utility Function

## 2.1 Definition and the Main Assumption

Let

$$F(x_1, x_2) \equiv \alpha \ln x_1 + \beta \ln x_2 - \gamma x_1 x_2.$$

The utility function we consider is defined as[4]

$$u(x_1, x_2) \equiv \begin{cases} F(x_1, x_2) & \text{for } (x_1, x_2) \in \mathbb{R}^2_+ - \Lambda_0, \\ \ln\left[(\alpha/\gamma)^{\alpha} x_2^{\beta-\alpha}\right] - \alpha & \text{for } (x_1, x_2) \in \Lambda_0, \end{cases}$$

where $\Lambda_0 \equiv \{(x_1, x_2) \in \mathbb{R}^2_+ : x_1 x_2 \geq \alpha/\gamma\}$. Parameters, $\alpha$, $\beta$, and $\gamma$ satisfy the following assumption.

**Assumption.** $\beta > \alpha > \frac{\beta}{2} > 0$ *and* $\gamma > 0$.

## 2.2 Properties

The function $u$ has all the standard properties as a utility function.

First, both $u$ and its partial derivatives with respect to $x_1$ and $x_2$ are continuous in $(x_1, x_2)$ over $\mathbb{R}^2_{++}$, as stated in the following Lemma.

**Lemma 1.** *For any* $(\tilde{x}_1, \tilde{x}_2) \in \mathbb{R}^2_{++}$,

$$\lim_{(x_1,x_2)\to(\tilde{x}_1,\tilde{x}_2)} u(x_1, x_2) = u(\tilde{x}_1, \tilde{x}_2),$$

$$\lim_{(x_1,x_2)\to(\tilde{x}_1,\tilde{x}_2)} u_i(x_1, x_2) = u_i(\tilde{x}_1, \tilde{x}_2),$$

*where* $u_i$ *is the partial derivative of* $u$ *with respect to* $x_i$, $i = 1, 2$.

[4]Even if the utility function takes the form of $F$ over $\mathbb{R}^2_+$, we can obtain all the following results except for global non-decreasingness and quasi-concavity of the utility function (Lemma 2).

**Fig. 1** Indifference curves

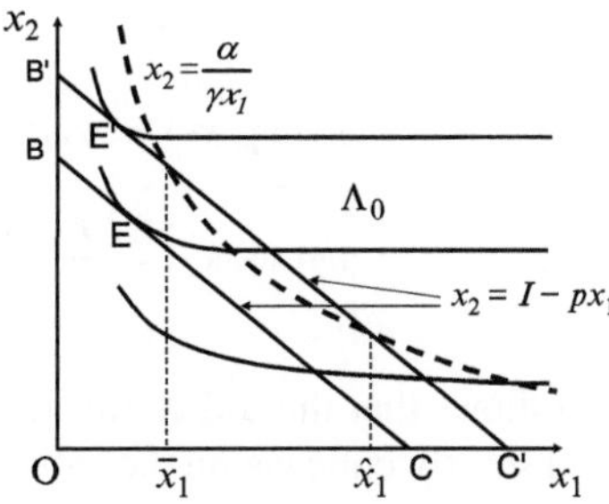


*Proof.* See Appendix. □

Second, we prove the following lemma, and then draw some indifference curves of the utility function (see Fig. 1). Note that the slope of any indifference curve is zero in $\Lambda_0$.

**Lemma 2.** *u is non-decreasing and quasi-concave in $x_1$ and $x_2$ over the domain $\mathbb{R}^2_{++}$.*

*Proof.* In fact, $u$ is concave in $x_1$ and $x_2$: For any $(x_1, x_2) \in \mathbb{R}^2_+ - \Lambda_0$,

$$u_1(x_1, x_2) = \frac{\alpha}{x_1} - \gamma x_2 > 0,$$

$$u_2(x_1, x_2) = \frac{\beta}{x_2} - \gamma x_1 > 0,$$

$$u_{11}(x_1, x_2) \equiv \frac{\partial^2}{\partial x_1^2} u(x_1, x_2) = -\frac{\alpha}{x_1^2} < 0,$$

$$u_{22}(x_1, x_2) \equiv \frac{\partial^2}{\partial x_2^2} u(x_1, x_2) = -\frac{\beta}{x_2^2} < 0,$$

$$u_{12}(x_1, x_2) \equiv \frac{\partial^2}{\partial x_2 \partial x_1} u(x_1, x_2) = -\gamma < 0,$$

$$\begin{aligned} \begin{vmatrix} u_{11} & u_{12} \\ u_{12} & u_{22} \end{vmatrix} &= \frac{\alpha\beta}{x_1^2 x_2^2} - \gamma^2 \\ &= \left[\frac{\sqrt{\alpha\beta}}{x_1 x_2} + \gamma\right]\left[\frac{\sqrt{\alpha\beta}}{x_1 x_2} - \gamma\right] \\ &> 0, \ (\because \alpha < \sqrt{\alpha\beta}, \text{ due to the Assumption}), \end{aligned}$$

and for any $(x_1, x_2) \in \Lambda_0$,

$$u_1(x_1, x_2) = u_{12}(x_1, x_2) = u_{11}(x_1, x_2) = 0,$$

$$u_2(x_1, x_2) = \frac{\beta - \alpha}{x_2} > 0, \quad u_{22}(x_1, x_2) = -\frac{\beta - \alpha}{x_2^2} < 0.$$

It follows that the utility function is quasi-concave over the commodity space $\mathbb{R}^2_+$ and strictly concave on the subset $\mathbb{R}^2_+ - \Lambda_0$. □

Third, the utility level on its border, $\alpha = \gamma x_1 x_2$, is

$$u\left(\frac{\alpha}{\gamma x_2}, x_2\right) = \ln\left[\left(\frac{\alpha}{\gamma}\right)^{\alpha} x_2^{\beta-\alpha}\right] - \alpha.$$

Due to the Assumption ($\beta > \alpha$), the utility level is monotonically increasing in $x_2$ along the hyperbola. Since $\lim_{x_2 \to 0} u\left(\frac{\alpha}{\gamma x_2}, x_2\right) = -\infty$ and $\lim_{x_2 \to \infty} u\left(\frac{\alpha}{\gamma x_2}, x_2\right) = \infty$, for any real number $\tilde{u}$, there is $(x_1, x_2)$ in $\mathbb{R}^2_{++}$ such that $\tilde{u} = u(x_1, x_2)$. In other words, $u$ is a mapping from $\mathbb{R}^2_{++}$ *onto* the set of real numbers $\mathbb{R}$.

Finally, considering that $u(0, x_2) = u(x_1, 0) = -\infty$, we see that any indifference curve cuts neither the horizontal nor vertical axis of coordinates; both goods are indispensable.

# 3 Derivation of the Demand Function

## 3.1 Utility Maximization Problem

Let $p$ and $I$ denote the price of good 1 and income, respectively. Good 2 serves as numeraire. Then, the utility maximization problem we consider is

$$\max_{\substack{x_1 \geq 0 \\ x_2 \geq 0}} u(x_1, x_2) \quad \text{subject to } px_1 + x_2 \leq I. \tag{1}$$

We show that for any positive price–income pair $(p, I)$, this problem has a unique, interior, and positive optimal consumption pair in $\mathbb{R}^2_{++} - \Lambda_0$.

First of all, let us define the function $U$ such that $U(x_1) \equiv u(x_1, I - px_1)$ for $0 \leq x_1 \leq I/p$. Then, the problem becomes[5]

$$\max_{0 \leq x_1 \leq I/p} U(x_1). \tag{2}$$

[5] Since the first derivatives $u_1$ and $u_2$ are nonnegative and positive, respectively, over $\mathbb{R}^2_+$, the budget constraint necessarily holds with equality.

Note that depending on the values of $p$ and $I$, we have two types of budget constraints. Let $G(x_1) \equiv \alpha - \gamma x_1(I - px_1)$.

Type I budget constraints fulfil the condition that $G(x_1) > 0$ for any $x_1 \in (0, I/p)$. An example is $BEC$ in Fig. 1.

Type II budget constraints are not Type I budget constraints; that is, each Type II budget constraint satisfies the following condition:

$$^{\exists}x_1 \in (0, I/p) \text{ such that } G(x_1) \leq 0.$$

An example is $B'E'C'$ in Fig. 1.

Since Type I (Type II) budget constraints imply that $I$ is relatively small (large) for a given $p$,[6] we may consider that Type I and Type II budget constraints refer to low and high income situations, respectively. Let us derive the optimal consumption pair for each type.

## 3.2 *Type I Budget Constraint*

Taking into account the definition of the utility function, we have

$$U(x_1) = \alpha \ln x_1 + \beta \ln(I - px_1) - \gamma x_1(I - px_1), \quad \text{for } x_1 \in [0, I/p].$$

Since

$$\lim_{x_1 \to 0} U(x_1) = \lim_{x_1 \to I/p} U(x_1) = -\infty,$$

there is an interior optimal solution to (2), which satisfies the first-order condition

$$\begin{aligned} U'(x_1) \equiv \frac{\mathrm{d}}{\mathrm{d}x_1} U(x_1) &= \frac{\alpha}{x_1} - \frac{\beta p}{I - px_1} - \gamma(I - 2px_1) \\ &= \left[\frac{\alpha}{x_1} - \gamma(I - px_1)\right] - p\left[\frac{\beta}{I - px_1} - \gamma x_1\right] \\ &= 0. \end{aligned} \tag{3}$$

To show the uniqueness of the optimal solution, let us prove the following lemma.

**Lemma 3.** *For any* $\tilde{x}_1 \in (0, I/p)$ *satisfying* $G(\tilde{x}_1) > 0$*, the second derivative of* $U$ *is negative, i.e.,*

[6]Since $G(x_1) = \gamma p\left(x_1 - \frac{I}{2p}\right)^2 + \alpha - \frac{\gamma I^2}{4p}$, Type I budget constraints correspond to the case of $I < 2(\alpha p/\gamma)^{\frac{1}{2}}$ and vice versa.

$$\left.\frac{\mathrm{d}^2}{\mathrm{d}x_1^2}U(x_1)\right|_{x_1=\bar{x}_1} < 0. \tag{4}$$

*Proof.* See Appendix. □

Lemma 3 implies that the second-order condition holds, and that the optimal solution to the maximization problem (2) is uniquely determined by the first-order condition for any given positive pair $(p, I)$ that generates a Type I budget constraint.

Note that the first-order condition (3) can be rewritten as the standard (price) = (the marginal rate of substitution) condition,

$$p = \frac{(I - px_1)[\alpha - \gamma x_1(I - px_1)]}{x_1[\beta - \gamma x_1(I - px_1)]}. \tag{5}$$

Point $E$ in Fig. 1 is the unique optimal solution corresponding to the Type I budget constraint $BEC$. This point satisfies (5).

### 3.3 Type II Budget Constraint

From the definition of $G(x_1)$, we find that $G(0) = G(I/p) = \alpha$. Since $G(I/2p) = \alpha - \gamma I^2/4p$ must be negative or zero in the case of Type II budget constraints, there exist $\bar{x}_1$ and $\hat{x}_1$ such that

$$0 < \bar{x}_1 \leq \hat{x}_1 < I/p$$

and

$$\begin{aligned} G(x_1) > 0 \quad &\text{for } x_1 \in (0, \bar{x}_1) \cup (\hat{x}_1, I/p), \\ G(x_1) < 0 \quad &\text{for } x_1 \in (\bar{x}_1, \hat{x}_1), \\ G(x_1) = 0 \quad &\text{for } x_1 = \bar{x}_1, \hat{x}_1. \end{aligned}$$

Based on this sign pattern of $G(x_1)$ and Lemma 1, which implies that both $U$ and its first derivative $U'$ are continuous in $x_1 \in (0, I/p)$, we are now in a position to prove the unique existence of the optimal solution to the maximization problem (2) for any given Type II budget constraint.

First, it follows from (3) that there exists $\epsilon > 0$ such that the first derivative, $U'(x_1)$, is positive for any $x_1 \in (0, \epsilon)$. Second, it is also clear from (3) that the first derivative is negative at $\bar{x}_1$ and $\hat{x}_1$. In fact, we can find that the first derivative is always negative in the interval $(\bar{x}_1, \hat{x}_1)$. Third, Lemma 3 ensures that $U'(x_1)$, too, is negative for any $x_1 > \hat{x}_1$. Since Lemma 3 also holds for the interval $(0, \bar{x}_1)$, we can conclude that the graph of $U(x_1)$ is bell-shaped with a single peak between 0 and $\bar{x}_1$, where the first-order condition

$$U'(x_1) = 0$$

is established. Point $E'$ in Fig. 1 is an example of the optimal solution.

### 3.4 First Main Proposition

Based on the foregoing argument, we can assert that whether the budget constraint is of Type I or Type II, the optimal solution is uniquely determined and differentiable with respect to $p$ and $I$. We now derive the first result of this paper.

**Proposition 1.** *For any positive income and price pair* $(p, I)$, *the utility maximization problem* (1) *has a unique, positive, and interior solution* $(\phi_1(p, I), \phi_2(p, I))$, *which naturally defines the demand functions for the goods:* $\phi_i : \mathbb{R}^2_{++} \to \mathbb{R}_+$, $i = 1, 2$. *Moreover, the demand functions are differentiable with respect to* $p$ *and* $I$.

## 4 Properties of the Demand Functions

To characterize the demand functions, $\phi_1$ and $\phi_2$, we focus on the budget constraint and the equality between the price and the marginal rate of substitution (5).

$$I = px_1 + x_2, \tag{6}$$

$$0 = (\beta - z)px_1 - (\alpha - z)x_2, \tag{7}$$

where $z \equiv \gamma x_1 x_2$.[7] Rearranging this system of equations yields

$$x_1 = \frac{(\alpha - z)I}{p(\alpha + \beta - 2z)}, \tag{8}$$

$$x_2 = \frac{(\beta - z)I}{\alpha + \beta - 2z}. \tag{9}$$

Combining (8) and (9), we have

$$\frac{p}{\gamma I^2} = \frac{(\alpha - z)(\beta - z)}{z(\alpha + \beta - 2z)^2} \equiv \Psi(z).$$

[7] As shown in the preceding section, for any $(p, I)$, the optimal consumption pair exists in $\mathbb{R}^2_{++} - \Lambda_0$; therefore, the value of $z$ that corresponds to the pair must be between 0 and $\alpha$.

**Fig. 2** The graph of $\Psi(z)$

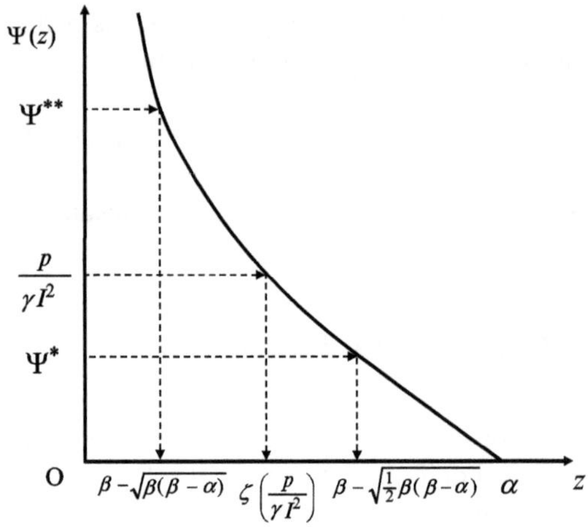


Figure 2 depicts the graph of $\Psi(z)$. As is clear from

$$\lim_{z \to 0+} \Psi(z) = \infty, \quad \Psi(\alpha) = 0, \tag{10}$$

and

$$\frac{\Psi'(z)z}{\Psi(z)} = -\left[1 + \frac{z(\alpha-\beta)^2}{(\alpha-z)(\beta-z)(\alpha+\beta-2z)}\right] < -1 \quad \text{for any } z \in (0,\alpha), \tag{11}$$

the value of $z$ that corresponds to the optimal consumption pair is uniquely determined for any positive $m \equiv p/\gamma I^2$.

Let us denote the inverse function of $\Psi$ by $\zeta$. It follows from (10) and (11) that

$$\lim_{m \to \infty} \zeta(m) = 0, \quad \zeta(0) = \alpha,$$

and

$$\frac{\zeta'(m)m}{\zeta(m)} = -\left[1 + \frac{\zeta(m)(\alpha-\beta)^2}{(\alpha-\zeta(m))(\beta-\zeta(m))(\alpha+\beta-2\zeta(m))}\right]^{-1}$$
$$\in (-1, 0) \quad \text{for any } m > 0. \tag{12}$$

Utilizing (8), (9), and the inverse function, we can express the demand functions as follows: For $(p, I) \in \mathbb{R}^2_{++}$,

$$\phi_1(p, I) = \frac{I\left(\alpha - \zeta\left(\frac{p}{\gamma I^2}\right)\right)}{p\left(\alpha + \beta - 2\zeta\left(\frac{p}{\gamma I^2}\right)\right)}, \tag{13}$$

$$\phi_2(p, I) = \frac{I\left(\beta - \zeta\left(\frac{p}{\gamma I^2}\right)\right)}{\alpha + \beta - 2\zeta\left(\frac{p}{\gamma I^2}\right)}. \tag{14}$$

In what follows, we shall use these expressions.[8]

## 4.1 Comparative Statics

Now that we have obtained the demand functions, we shall determine how the demand for good 1 and good 2 depends on $p$ and $I$.

### 4.1.1 Price Effects

First, the demand for good 1 has a distinctive feature such that the sign of $\partial\phi_1(p, I)/\partial p$ depends solely on the value of $\zeta(m)$, as proved in the following lemma.

**Lemma 4.1.** *For any given price–income pair,* (i) $\partial\phi_1(p, I)/\partial p$ *is negative if* $\zeta(p/\gamma I^2) \in \left(0, \beta - \sqrt{\beta(\beta-\alpha)/2}\right)$*;* (ii) $\partial\phi_1(p, I)/\partial p$ *is positive if* $\zeta(p/\gamma I^2) \in \left(\beta - \sqrt{\beta(\beta-\alpha)/2}, \alpha\right)$.

*Proof.* The logarithmic differentiation of $\phi_1$ with respect to $p$ yields

$$\frac{p}{\phi_1(p, I)} \cdot \frac{\partial\phi_1(p, I)}{\partial p} = -1 + \frac{\zeta(m)(\alpha - \beta)}{(\alpha - \zeta(m))(\alpha + \beta - 2\zeta(m))} \cdot \frac{\zeta'(m)m}{\zeta(m)}.$$

Utilizing (12), we derive

$$\frac{p}{\phi_1(p, I)} \cdot \frac{\partial\phi_1(p, I)}{\partial p} \tag{15}$$

$$= \frac{2(\zeta(m) - \alpha)\left[\zeta(m) - \left\{\beta - \sqrt{\frac{1}{2}\beta(\beta-\alpha)}\right\}\right]\left[\zeta(m) - \left\{\beta + \sqrt{\frac{1}{2}\beta(\beta-\alpha)}\right\}\right]}{\zeta(m)(\alpha-\beta)^2 + (\alpha - \zeta(m))(\beta - \zeta(m))(\alpha + \beta - 2\zeta(m))},$$

where the denominator is positive, due to $\beta > \alpha > \zeta(m)$ for any positive $m$.

[8]Needless to say, from the definition of $\zeta$, the equality $\zeta(p/\gamma I^2) = \gamma\phi_1(p, I)\phi_2(p, I)$ holds.

Let us denote the numerator as

$$\begin{aligned}&\Xi(\zeta(m))\\&\equiv 2\,(\zeta(m)-\alpha)\left[\zeta(m)-\left\{\beta-\sqrt{\tfrac{1}{2}\beta(\beta-\alpha)}\right\}\right]\left[\zeta(m)-\left\{\beta+\sqrt{\tfrac{1}{2}\beta(\beta-\alpha)}\right\}\right].\end{aligned}$$

It is clear that $\Xi(0)=-\alpha\beta(\alpha+\beta)<0$. Further, we see that the inequality

$$\alpha>\beta-\sqrt{\tfrac{1}{2}\beta(\beta-\alpha)}$$

holds, due to the Assumption.[9] Considering that $\Xi(0)<0$, we derive

$$\begin{aligned}\Xi(\zeta(m))<0 \quad &\text{for } \zeta(m)\in\left(0,\beta-\sqrt{\tfrac{1}{2}\beta(\beta-\alpha)}\right),\\ \Xi(\zeta(m))>0 \quad &\text{for } \zeta(m)\in\left(\beta-\sqrt{\tfrac{1}{2}\beta(\beta-\alpha)},\alpha\right),\end{aligned}$$

and (15) implies

$$\operatorname{sign}\left[\frac{\partial\phi_1(p,I)}{\partial p}\right]=\operatorname{sign}\left[\Xi\left(\zeta\left(\frac{p}{\gamma I^2}\right)\right)\right].$$

Therefore, Lemma 4.1 is established. □

Second, let us partially differentiate $\phi_2$ with respect to $p$. From the partial differentiation of (14), we obtain

$$\frac{\partial\phi_2(p,I)}{\partial p}=\frac{(\beta-\alpha)\zeta'(m)}{\gamma I(\alpha+\beta-2\zeta(m))^2}<0.$$

Finally, since $\zeta$ is the inverse function of $\Psi$, from Lemma 4.1, we obtain the following result.

**Proposition 2.** (i) *If a pair* $(p,I)$ *satisfies* $\Psi^*\equiv\Psi\big(\beta-\sqrt{\beta(\beta-\alpha)/2}\big)<p/\gamma I^2$, *then* $\partial\phi_1(p,I)/\partial p$ *is negative for the pair;* (ii) *If they satisfy* $\Psi^*>p/\gamma I^2>0$, *then* $\partial\phi_1(p,I)/\partial p$ *is positive for the pair; good* 1 *is a Giffen good at the price and income levels;* (iii) $\partial\phi_2(p,I)/\partial p$ *is always negative.*

---

[9] $(\beta-\alpha)-\sqrt{\beta(\beta-\alpha)/2}<0$ holds for

$$\begin{aligned}\left[(\beta-\alpha)-\sqrt{\tfrac{1}{2}\beta(\beta-\alpha)}\right]\times\left[(\beta-\alpha)+\sqrt{\tfrac{1}{2}\beta(\beta-\alpha)}\right]&=(\beta-\alpha)^2-\frac{\beta(\beta-\alpha)}{2}\\&=(\beta-\alpha)\left(\frac{\beta}{2}-\alpha\right),\end{aligned}$$

which is negative due to the Assumption.

### 4.1.2 Income Effects

Next, let us investigate the income effects on the demand for good 1 and good 2. Similar to the case of price effects, we first derive the following lemma.

**Lemma 4.2.** *For any given price–income pair,* (i) $\partial\phi_1(p, I)/\partial I$ *is positive if* $\zeta(p/\gamma I^2) \in \left(0, \beta - \sqrt{\beta(\beta-\alpha)}\right)$; (ii) $\partial\phi_1(p, I)/\partial I$ *is negative if* $\zeta(p/\gamma I^2) \in \left(\beta - \sqrt{\beta(\beta-\alpha)}, \alpha\right)$.

*Proof.* See Appendix. □

Second, let us partially differentiate $\phi_2$ with respect to $I$. From the partial differentiation of (14), we obtain

$$\frac{\partial\phi_2(p, I)}{\partial I} = \frac{\beta - \zeta(m)}{\alpha + \beta - 2\zeta(m)} - \frac{2p(\beta-\alpha)\zeta'(m)}{\gamma I^2(\alpha+\beta-2\zeta(m))^2} > 0.$$

From Lemma 4.2, we obtain the following result.

**Proposition 3.** (i) *If a pair* $(p, I)$ *satisfies* $\Psi^{**} \equiv \Psi\left(\beta - \sqrt{\beta(\beta-\alpha)}\right) < p/\gamma I^2$, *then* $\partial\phi_1(p, I)/\partial I$ *is positive for the pair; good* 1 *is a normal good at the price and income levels;* (ii) *If they satisfy* $\Psi^{**} > p/\gamma I^2 > 0$, *then* $\partial\phi_1(p, I)/\partial I$ *is negative for the pair; good* 1 *is an inferior good at the price and income levels;* (iii) $\partial\phi_2(p, I)/\partial I$ *is always positive.*

### 4.1.3 Diagrammatic Explanation of the Comparative Static Analysis

Let us explain Propositions 2 and 3 diagrammatically, and then depict the income expansion path and the price offer curve. See Fig. 3a, where the horizontal and vertical axes of coordinates measure $p$ and $I$, respectively; curve *OGC* and *OAH* are two quadratic curves $\Psi^* = p/\gamma I^2$ and $\Psi^{**} = p/\gamma I^2$. The propositions above imply that (i) good 1 is a Giffen good at $(p, I)$ in Region I, (ii) good 1 is an inferior good with a negatively sloped demand curve at $(p, I)$ in Region II and is a normal good at $(p, I)$ in Region III.

The signs of $\partial\phi_1(p, I)/\partial p$ and $\partial\phi_1(p, I)/\partial I$ depend on whether the value $\gamma\phi_1(p, I)\phi_2(p, I)$, which is uniquely determined by $p/\gamma I^2$, is greater or smaller than $\beta - \sqrt{\beta(\beta-\alpha)/2}$ and $\beta - \sqrt{\beta(\beta-\alpha)}$ (Lemmas 4.1, 4.2). Thus, as is shown in Fig. 3b, the commodity space is divided by three hyperbolas:

$$\gamma x_1 x_2 = \alpha,$$

$$\gamma x_1 x_2 = \beta - \sqrt{\tfrac{1}{2}\beta(\beta-\alpha)},$$

$$\gamma x_1 x_2 = \beta - \sqrt{\beta(\beta-\alpha)}.$$

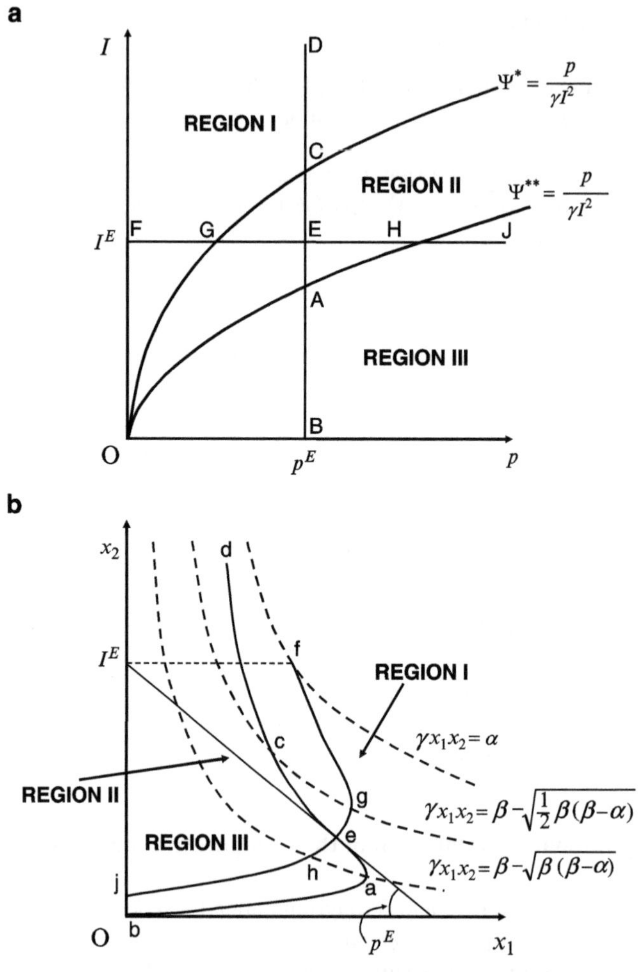


**Fig. 3** (**a**) The price–income space. (**b**) The income expansion path and the price offer curve

Regions I, II, and III in Fig. 3b correspond to Regions I, II, and III, respectively, in Fig. 3a. Take any positive point in the commodity space, say $(x_1^e, x_2^e)$. There is an indifference curve and its tangent line at point $e$; the tangent line is expressed as

$$\frac{u_1\left(x_1^e, x_2^e\right)}{u_2\left(x_1^e, x_2^e\right)} x_1 + x_2 = \frac{u_1\left(x_1^e, x_2^e\right)}{u_2\left(x_1^e, x_2^e\right)} x_1^e + x_2^e.$$

The pair $(x_1^e, x_2^e)$ is the solution to the maximization problem (1) if the price is $p^E \equiv \frac{u_1(x_1^e, x_2^e)}{u_2(x_1^e, x_2^e)}$ and the income is $I^E \equiv \frac{u_1(x_1^e, x_2^e)}{u_2(x_1^e, x_2^e)} x_1^e + x_2^e$; we have $x_1^e = \phi_1(p^E, I^E)$ and $x_2^e = \phi_2(p^E, I^E)$. For example, suppose that point $e$ is in Region II in Fig. 3b. Then, $p^E/\gamma(I^E)^2$ is in between $\Psi^*$ and $\Psi^{**}$; point $(p^E, I^E)$ has to be in Region II in Fig. 3a, in which case good 1 is an inferior good with a negatively sloped demand curve at $(p^E, I^E)$, i.e., both $\partial\phi_1(p^E, I^E)/\partial p$ and $\partial\phi_1(p^E, I^E)/\partial I$ are negative. By this correspondence, the points in Fig. 3a, b have a one-to-one relationship.

### 4.1.4 Income Expansion Path and Price Offer Curve

Based on the foregoing argument, we can draw the income expansion path and the price offer curve,[10] both of which commonly cross a point, say $E$, in Fig. 3a. Point $E$ in Fig. 3a corresponds to point $e$ in Fig. 3b, and the vertical line *BAECD* in Fig. 3a corresponds to the income expansion path *baecd* in Fig. 3b; further, the horizontal line *JHEGF* in Fig. 3a corresponds to the price offer curve *jhegf* in Fig. 3b. The income expansion path and the price offer curve bend when they cross the hyperbolas $\beta - \sqrt{\beta(\beta-\alpha)} = \gamma x_1 x_2$ and $\beta - \sqrt{\beta(\beta-\alpha)/2} = \gamma x_1 x_2$, respectively. One can also verify that the Engel curve of good 1 is bell-shaped.

Moving up along the vertical line *BAECD* in Fig. 3a, the ratio $p/\gamma I^2$ monotonically decreases, which means that $\zeta(p/\gamma I^2)$ $(= \gamma\phi_1(p, I)\phi_2(p, I))$ monotonically increases. The path rises between the hyperbola $\alpha = \gamma x_1 x_2$ and the vertical axis of coordinates.

On the other hand, moving in the right direction along the horizontal line *FGEHJ* in Fig. 3a, the ratio $p/\gamma I^2$ monotonically increases, which means that $\zeta(p/\gamma I^2)$ $(= \gamma\phi_1(p, I)\phi_2(p, I))$ monotonically decreases. The price offer curve converges to a point on the vertical axis of coordinates, say $j$ in Fig. 3b. Note that because of (14), $\lim\limits_{\zeta(p/\gamma I^2)\to 0} \phi_2(p, I) = I\beta/(\alpha+\beta)$.

### 4.1.5 Independence of Giffen Behavior from Living Standard

For an arbitrarily chosen hyperbola $\tilde{z} = \gamma x_1 x_2$ $(0 < \tilde{z} \leq \alpha)$, the utility level on it is given by

$$u\left(\frac{\tilde{z}}{\gamma x_2}, x_2\right) = \ln\left[\left(\frac{\tilde{z}}{\gamma}\right)^{\alpha} x_2^{\beta-\alpha}\right] - \tilde{z},$$

which monotonously increases from $-\infty$ to $+\infty$, as $x_2$ increases from 0 to $+\infty$. Therefore, any indifference curve crosses the hyperbola $\tilde{z} = \gamma x_1 x_2$, which obviously means that it has to cross the two border curves $\beta - \sqrt{\beta(\beta-\alpha)} = \gamma x_1 x_2$ and $\beta - \sqrt{\beta(\beta-\alpha)/2} = \gamma x_1 x_2$. Based on this fact, we derive the following result.

[10] The definitions of income expansion path and price offer curve are given in standard textbooks of microeconomic theory. For example, see Varian [10, p. 116–118].

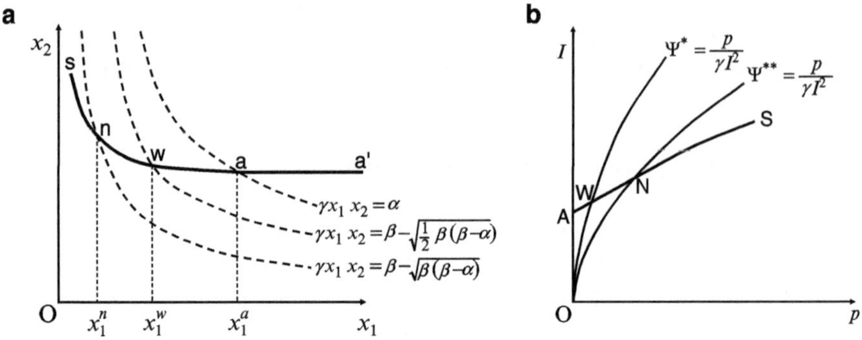


**Fig. 4** (**a**) The three portions of an indifference curve. (**b**) The indifference curve in the parameter space

**Proposition 4.** *Define a set of positive pairs* $(p, I)$

$$\Omega(\tilde{u}) \equiv \left\{(p, I) \in \mathbb{R}^2_{++} : \tilde{u} = u(\phi_1(p, I), \phi_2(p, I))\right\}.$$

*For any utility level* $\tilde{u}$*, there are three nonempty subsets* $\Omega_g(\tilde{u})$*,* $\Omega_i(\tilde{u})$*, and* $\Omega_n(\tilde{u})$ *such that* (i) *for any* $(p, I) \in \Omega_g(\tilde{u})$*,*

$$\frac{\partial \phi_1(p, I)}{\partial p} > 0 \quad and \quad \frac{\partial \phi_1(p, I)}{\partial I} < 0;$$

*good* 1 *is a Giffen good at* $(p, I) \in \Omega_g(\tilde{u})$*;* (ii) *for any* $(p, I) \in \Omega_i(\tilde{u})$*,*

$$\frac{\partial \phi_1(p, I)}{\partial p} < 0 \quad and \quad \frac{\partial \phi_1(p, I)}{\partial I} < 0;$$

*good* 1 *is an inferior good with a negatively sloped demand curve at* $(p, I) \in \Omega_i(\tilde{u})$*;* (iii) *for any* $(p, I) \in \Omega_n(\tilde{u})$*,*

$$\frac{\partial \phi_1(p, I)}{\partial p} < 0 \quad and \quad \frac{\partial \phi_1(p, I)}{\partial I} > 0;$$

*good* 1 *is a normal good at* $(p, I) \in \Omega_n(\tilde{u})$.

*Proof.* See Fig. 4a, where $a'awns$ is an arbitrarily chosen indifference curve. As we proved prior to Proposition 4, the curve has to cross the two border hyperbolas at $w$ and $n$. Corresponding to the indifference curve $awns$ in Fig. 4a, we can derive the curve $AWNS$ in Fig. 4b; points $w$ and $n$ in Fig. 4a correspond to points $W$ and $N$, respectively, in Fig. 4b. As is clear from what we have shown so far, good 1 is a Giffen good when $(\phi_1(p, I), \phi_2(p, I))$ is on $aw$, an inferior good with a negatively sloped demand curve when it is on $wn$, and a normal good when it is on $ns$. Therefore, we can define $\Omega_g(\tilde{u})$, $\Omega_i(\tilde{u})$ and $\Omega_n(\tilde{u})$ as follows:

$$\begin{aligned}\Omega_g(\tilde{u}) &= \{(p,I) \in \mathbb{R}^2_{++} : (p,I) \text{ is on } AW \text{ in Fig. 4b}\}\\ \Omega_i(\tilde{u}) &= \{(p,I) \in \mathbb{R}^2_{++} : (p,I) \text{ is on } WN \text{ in Fig. 4b}\}\\ \Omega_n(\tilde{u}) &= \{(p,I) \in \mathbb{R}^2_{++} : (p,I) \text{ is on } NS \text{ in Fig. 4b}\}\,. \qquad \square\end{aligned}$$

*Remark 1.* Formally, the curve *AWNS* in Fig. 4b is expressed as follows. Let $x_2 = \varsigma(x_1;\tilde{u})$ be an indifference curve with the utility level $\tilde{u}$. Then, the corresponding $p$ and $I$ are expressed as

$$p(x_1) \equiv -\left.\frac{\partial\varsigma(s;\tilde{u})}{\partial s}\right|_{s=x_1},$$

$$I(x_1) \equiv \left(-\left.\frac{\partial\varsigma(s;\tilde{u})}{\partial s}\right|_{s=x_1}\right)x_1 + \varsigma(x_1;\tilde{u}).$$

Any point on the curve *AWNS* in Fig. 4b is expressed as $(p(x_1), I(x_1))$. For example, point $A$ is $(p(x_1^a), I(x_1^a))$, where $p(x_1^a) = 0$. Similarly, point $W$ and point $N$ are $(p(x_1^w), I(x_1^w))$ and $(p(x_1^n), I(x_1^n))$, respectively.

### 4.1.6 Expenditure Share on a Giffen Good

Suppose that the consumption pair $(x_1, x_2)$ is optimally chosen for a given $(p, I)$; then, from (6) and (7), the share of income spent on good 1 is given by

$$\frac{px_1}{I} = \frac{\frac{(\alpha-z)x_2}{(\beta-z)x_1}x_1}{\frac{(\alpha-z)x_2}{(\beta-z)x_1}x_1 + x_2} = \frac{\alpha - z}{\alpha+\beta-2z}.$$

Let $\vartheta(z) \equiv \frac{\alpha-z}{\alpha+\beta-2z}$; then,

$$\vartheta(z) \in \left(0, \frac{\alpha}{\alpha+\beta}\right) \quad \text{and} \quad \vartheta'(z) = \frac{\alpha-\beta}{(\alpha+\beta-2z)^2} < 0 \qquad \text{for } z \in (0,\alpha).(16)$$

Based on the above, we obtain the following result.

**Proposition 5.** *Denote a share of income spent on good 1 by $\theta$, that is, $\theta \equiv p\phi_1(p,I)/I$. There exist the following relations between the property of demand for good* 1 *and $\theta$.* (i) *Good* 1 *is a Giffen good if $\theta \in (0,\theta_g)$;* (ii) *Good* 1 *is an inferior good with a negatively sloped demand curve if $\theta \in (\theta_g, \theta_i)$;* (iii) *Good* 1 *is a normal good if $\theta \in \left(\theta_i, \frac{\alpha}{\alpha+\beta}\right)$, where $\theta_g$ and $\theta_i$ are defined as follows.*

$$\theta_g \equiv \frac{\alpha-\beta+\sqrt{\frac{1}{2}\beta(\beta-\alpha)}}{\alpha-\beta+\sqrt{2\beta(\beta-\alpha)}},$$

$$\theta_i \equiv \frac{\alpha-\beta+\sqrt{\beta(\beta-\alpha)}}{\alpha-\beta+2\sqrt{\beta(\beta-\alpha)}}.$$

*Proof.* Considering the fact that for any $(p, I)$, the expenditure share on good 1 is given by $\vartheta(\zeta(p/\gamma I^2))$, from (16), Lemma 4.1, and Lemma 4.2, we can conclude that

$$\frac{\partial \phi_1(p, I)}{\partial p} > 0 \quad \text{and} \quad \frac{\partial \phi_1(p, I)}{\partial I} < 0$$

$$\text{if } \theta \in \left(\vartheta(\alpha), \vartheta\left(\beta - \sqrt{\tfrac{1}{2}\beta(\beta - \alpha)}\right)\right),$$

$$\frac{\partial \phi_1(p, I)}{\partial p} < 0 \quad \text{and} \quad \frac{\partial \phi_1(p, I)}{\partial I} < 0$$

$$\text{if } \theta \in \left(\vartheta\left(\beta - \sqrt{\tfrac{1}{2}\beta(\beta - \alpha)}\right), \vartheta\left(\beta - \sqrt{\beta(\beta - \alpha)}\right)\right),$$

$$\frac{\partial \phi_1(p, I)}{\partial p} < 0 \quad \text{and} \quad \frac{\partial \phi_1(p, I)}{\partial I} > 0$$

$$\text{if } \theta \in \left(\vartheta\left(\beta - \sqrt{\beta(\beta - \alpha)}\right), \vartheta(0)\right),$$

where $\vartheta(\alpha) = 0$, $\vartheta\left(\beta - \sqrt{\beta(\beta - \alpha)/2}\right) = \theta_g$, $\vartheta\left(\beta - \sqrt{\beta(\beta - \alpha)}\right) = \theta_i$, and $\vartheta(0) = \alpha/(\alpha + \beta)$. □

Let us consider the change along an income expansion path. As households' income $I$ goes up, which means that $\zeta(p/\gamma I^2)$ monotonously rises, their expenditure share on good 1 monotonously declines, but at the same time, good 1 becomes a Giffen good. However low the expenditure share on good 1 becomes, it remains a Giffen good.

# 5 Possible Generalizations

Finally, we consider some possible generalizations of the utility function proposed in this paper. First, we can slightly modify the utility function in such a way that it becomes increasing and strictly quasi-concave in $x_1$ and $x_2$ (see Fig. 5). The curve *ABCD* is an arbitrarily chosen indifference curve of the utility function. Choose a small positive number $\varepsilon$ that satisfies $\alpha - \left[\beta - \sqrt{\beta(\beta - \alpha)/2}\right] > \varepsilon$. We can draw a curve *BFG* such that (i) it is tangent to the original indifference curve at $B$, (ii) it is a convex curve, and (iii) it asymptotically converges to the horizontal part of the original indifference curve *ABCD*. Define *ABFG* as a modified indifference curve. Making such modifications to each indifference curve, we can derive a modified utility function that is increasing and strictly quasi-concave in $x_1$ and $x_2$.

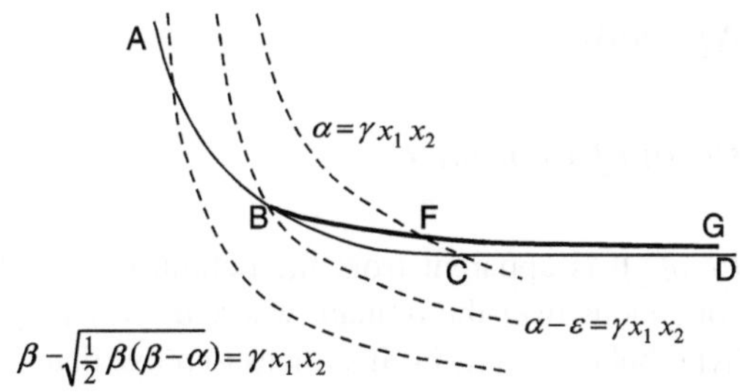


**Fig. 5** Strictly quasi-concave utility function

Second, one may wonder whether there is a family of utility functions that has the same properties as the function $u$. The answer is affirmative. Iwasa and Shimomura [2] show that one can define a non-decreasing, globally quasi-concave utility function that generates Giffen behavior by partially using the following functional form:

$$f(x_1, x_2) \equiv \frac{\alpha}{1-\sigma}\left(x_1^{1-\sigma}-1\right)+\frac{\beta}{1-\eta}\left(x_2^{1-\eta}-1\right)-\gamma x_1 x_2,$$

where all the parameters are positive. Clearly, the form of $f$ converges to that of $F$ as both $\sigma$ and $\eta$ go to 1. The main purpose of their paper is the same as that of the early studies, that is, to present examples of the utility function with the Giffen property, and hence, it lacks the comprehensive analysis that this paper has.

## 6 Concluding Remarks

In this paper, we have proposed a specific but standard utility function under which one good can be an inferior good or even a Giffen good. We show that Giffen behavior generated from the utility function is compatible with an arbitrarily high level of utility and low share of income spent on the inferior good. Thus, it is far removed from the interpretation of the Giffen paradox commonly found in textbooks and reveals the theoretical possibility that it is inappropriate to presume that the Giffen paradox is merely an exceptional phenomenon that occurs only under extreme circumstances.

We believe that there are some research fields in which this type of utility function plays an important role. Let us conclude this paper by providing an example. Using the utility function $u$ or its variant, Nishimura and Shimomura [6] and Doi et al. [1] show that a two-country dynamic general equilibrium model of international trade can generate local indeterminacy even if externalities, public goods, imperfect competition, or government intervention are assumed away. These papers show that negative income effects are crucial in order to obtain local indeterminacy.

# Appendix

## Proof of Lemma 1

*Proof.* It is apparent from the definition of $u$ that both $u$ and $u_i$ $(i = 1, 2)$ are continuous over the domain where $\alpha \neq \gamma x_1 x_2$. Let us focus on any point of the hyperbola $\alpha = \gamma \tilde{x}_1 \tilde{x}_2$. It suffices to prove that

1. $\lim_{(x_1,x_2)\to(\tilde{x}_1,\tilde{x}_2)} F(x_1, x_2) = \ln\left[(\alpha/\gamma)^\alpha \tilde{x}_2^{\beta-\alpha}\right] - \alpha \;\{= u(\tilde{x}_1, \tilde{x}_2)\}$,
2. $\lim_{(x_1,x_2)\to(\tilde{x}_1,\tilde{x}_2)} F_1(x_1, x_2) = 0 \left\{= \frac{\partial}{\partial x_1}\left(\ln\left[(\alpha/\gamma)^\alpha x_2^{\beta-\alpha}\right] - \alpha\right)\Big|_{(x_1,x_2)=(\tilde{x}_1,\tilde{x}_2)}\right\}$,
3. $\lim_{(x_1,x_2)\to(\tilde{x}_1,\tilde{x}_2)} F_2(x_1, x_2) = \frac{\beta-\alpha}{\tilde{x}_2} \left\{= \frac{\partial}{\partial x_2}\left(\ln\left[(\alpha/\gamma)^\alpha x_2^{\beta-\alpha}\right] - \alpha\right)\Big|_{(x_1,x_2)=(\tilde{x}_1,\tilde{x}_2)}\right\}$,

where $F_i$ is the partial derivative of $F$ with respect to $x_i$, $i = 1, 2$.

Let us prove 1 here. The others can be proved in a similar manner.

Since

$$
\begin{aligned}
&F(x_1, x_2) - \left(\ln\left[(\alpha/\gamma)^\alpha \tilde{x}_2^{\beta-\alpha}\right] - \alpha\right)\\
&= \alpha \ln x_1 + \beta \ln x_2 - \gamma x_1 x_2 - \left[\alpha \ln \frac{\alpha}{\gamma \tilde{x}_2} + \beta \ln \tilde{x}_2 - \alpha\right]\\
&= \alpha \ln \frac{\gamma x_1 \tilde{x}_2}{\alpha} + \beta \ln \frac{x_2}{\tilde{x}_2} - (\gamma x_1 x_2 - \alpha),
\end{aligned}
$$

it is apparent from the continuity of $\gamma x_1 x_2$ that for any $\delta > 0$, there is some $\epsilon > 0$ such that

$$
\max\left\{|x_1 - \tilde{x}_1|, |x_2 - \tilde{x}_2|\right\} < \epsilon \Rightarrow \left|F(x_1, x_2) - \left(\ln\left[(\alpha/\gamma)^\alpha \tilde{x}_2^{\beta-\alpha}\right] - \alpha\right)\right| < \delta.
$$

Therefore, 1 is established. □

## Proof of Lemma 3

*Proof.* Considering (3), the differentiation of the first derivative at $\tilde{x}_1$ yields

$$
\begin{aligned}
\frac{d^2}{dx_1^2} U(x_1)\Big|_{x_1=\tilde{x}_1} &= -\left[\frac{\beta p^2}{(I - p\tilde{x}_1)^2} - 2\gamma p + \frac{\alpha}{\tilde{x}_1^2}\right]\\
&= -\beta\left[\frac{p}{(I - p\tilde{x}_1)} - \frac{\gamma}{\beta}(I - p\tilde{x}_1)\right]^2 - \frac{(I - p\tilde{x}_1)^2}{\beta}\left[\frac{\beta\alpha}{\tilde{x}_1^2(I - p\tilde{x}_1)^2} - \gamma^2\right].
\end{aligned}
$$

Here, the term $[\frac{\beta\alpha}{\tilde{x}_1^2(I-p\tilde{x}_1)^2} - \gamma^2]$ is positive if $G(\tilde{x}_1) = \alpha - \gamma\tilde{x}_1(I - p\tilde{x}_1) > 0$ and $\tilde{x}_1 \in (0, I/p)$. For

$$\begin{aligned}
\frac{\beta\alpha}{\tilde{x}_1^2(I-p\tilde{x}_1)^2} - \gamma^2 &= \frac{\left[\frac{\sqrt{\beta\alpha}}{\tilde{x}_1} + \gamma(I - p\tilde{x}_1)\right]\left[\frac{\sqrt{\beta\alpha}}{\tilde{x}_1} - \gamma(I - p\tilde{x}_1)\right]}{(I-p\tilde{x}_1)^2} \\
&> \frac{\left[\frac{\sqrt{\beta\alpha}}{\tilde{x}_1} + \gamma(I - p\tilde{x}_1)\right]\left[\frac{\alpha}{\tilde{x}_1} - \gamma(I - p\tilde{x}_1)\right]}{(I-p\tilde{x}_1)^2} \\
&> 0,
\end{aligned}$$

where the first inequality comes from $\alpha < \sqrt{\beta\alpha}$ ($\because$ Assumption) and the second inequality is due to $G(\tilde{x_1}) > 0$. Therefore, (4) is established for any $\tilde{x}_1 \in (0, I/p)$ if $G(\tilde{x}_1) > 0$. □

## *Proof of Lemma 4.2*

*Proof.* The logarithmic differentiation of $\phi_1$ with respect to $I$ yields

$$\begin{aligned}
&\frac{I}{\phi_1(p,I)} \cdot \frac{\partial \phi_1(p,I)}{\partial I} \\
&= 1 + \frac{2\zeta(m)(\beta-\alpha)}{(\alpha-\zeta(m))(\alpha+\beta-2\zeta(m))} \cdot \frac{\zeta'(m)m}{\zeta(m)} \\
&= 1 - \frac{2\zeta(m)(\beta-\alpha)(\beta-\zeta(m))}{\zeta(m)(\alpha-\beta)^2 + (\alpha-\zeta(m))(\beta-\zeta(m))(\alpha+\beta-2\zeta(m))} \\
&= \frac{\left[\alpha\beta - 2\beta\zeta(m) + \zeta(m)^2\right](\alpha+\beta-2\zeta(m))}{\zeta(m)(\alpha-\beta)^2 + (\alpha-\zeta(m))(\beta-\zeta(m))(\alpha+\beta-2\zeta(m))}.
\end{aligned}$$

Let

$$\Upsilon(\zeta(m)) \equiv \alpha\beta - 2\beta\zeta(m) + \zeta(m)^2.$$

The sign of $\partial\phi_1(p,I)/\partial I$ is equal to that of $\Upsilon(\zeta(m))$. Since $\Upsilon(0) = \alpha\beta > 0$ and $\Upsilon(\alpha) = -\alpha(\beta-\alpha) < 0$, we derive

$$\text{sign}\left[\frac{\partial\phi_1(p,I)}{\partial I}\right] = \text{sign}\left[\Upsilon\left(\zeta\left(\frac{p}{\gamma I^2}\right)\right)\right] > 0$$

$$\text{if} \quad \zeta\left(\frac{p}{\gamma I^2}\right) \text{ is in the interval } \left(0, \beta - \sqrt{\beta(\beta-\alpha)}\right)$$

and

$$\operatorname{sign}\left[\frac{\partial \phi_1(p,I)}{\partial I}\right] = \operatorname{sign}\left[\Upsilon\left(\zeta\left(\frac{p}{\gamma I^2}\right)\right)\right] < 0$$

$$\text{if} \quad \zeta\left(\frac{p}{\gamma I^2}\right) \text{ is in the interval } \left(\beta - \sqrt{\beta(\beta-\alpha)}, \alpha\right).$$

Therefore, Lemma 4.2 is established. □

**Acknowledgements** Comments from Murray C. Kemp, Binh Tran-Nam, Ngo Van Long, Masao Oda, Noritsugu Nakanishi, and Chiaki Hara have greatly improved the paper. We have also benefited from discussions with Koichi Hamada, Satya Das, Takashi Kamihigashi, Tomoyuki Kamo, Toru Kikuchi, Katsufumi Fukuda, Yu-chin Chen, Fahad Khalil, Takeshi Nakatani, Kazuo Nishimura, Ken-Ichi Shimomura, and Stephen J. Turnovsky. We thank the anonymous referee for his/her helpful comments. Iwasa would like to acknowledge financial support from Research Fellowships of the Japan Society for the Promotion of Science for Young Scientists. Professor Koji Shimomura passed away on February 24, 2007. This paper was completed after his untimely death.

## References

1. Doi, J., Iwasa, K., Shimomura, K.: Indeterminacy in the free-trade world. J Differ Equ Appl **13**, 135–149 (2007).
2. Iwasa, K., Shimomura, K.: A family of utility functions which generate Giffen paradox. J Econ Kwansei Gakuin Univ **60**(3), 29–45 (2007).
3. Mas-Colell, A., Whinston, M.D., Green, J.R.: Microeconomic Theory. Oxford: Oxford University Press (1995).
4. Moffatt, P.G.: Is Giffen behavior compatible with the axioms of consumer theory? J Math Econ **37**, 259–267 (2002).
5. Nachbar, J.H.: The last word on Giffen goods? Econ Theory **11**, 403–412 (1998).
6. Nishimura, K., Shimomura, K.: Indeterminacy in a dynamic two-country model. Econ Theory **29**, 307–324 (2006).
7. Sørensen, P.N.: Simple utility functions with Giffen demand. Econ Theory **31**, 367–370 (2007).
8. Spiegel, U.: The case of a Giffen good. J Econ Educ **25**, 137–147 (1994).
9. Vandermeulen, D.C.: Upward sloping demand curve without the Giffen paradox. Am Econ Rev **72**, 453–458 (1972).
10. Varian, H.R.: Microeconomic Analysis, 3rd edn. New York: W.W. Norton & Company (1991).
11. Weber, C.E.: The case of a Giffen good: Comment. J Econ Educ **28**, 36–44 (1997).
12. Wold, H., Jureen, L.: Demand Analysis. New York: Wiley (1953).

# A Class of Indirect Utility Functions Predicting Giffen Behaviour

**Peter G. Moffatt**

**Abstract** The problem of recognising Giffen behaviour is approached from the standpoint of the indirect utility function (IUF) from which the Marshallian demands are easily obtained via Roy's identity. It is shown that, for the two-good situation, downward convergence of the contours of the IUF is necessary for giffenity, and sufficient if this downward convergence is strong enough, in a sense that is geometrically determined. A family of IUFs involving hyperbolic contours convex to the origin, and having this property of (locally) downward convergence is constructed. The Marshallian demands are obtained, and the region of Giffen behaviour determined. For this family, such regions exist for each good, and are non-overlapping. Finally, it is shown by geometric construction that the family of Direct Utility Functions having the same pattern of hyperbolic contours also exhibits giffenity in corresponding subregions of the positive quadrant.

## 1 Introduction

Ever since Wold and Jureen's [8] attempt to illustrate the Giffen paradox by specifying a particular direct utility function, there has been a stream of contributions from authors pursuing similar objectives, for example Spiegel [4], Weber [6, 7], Moffatt [1] and Sørensen [3].

One of the lessons learned from this strand of literature is that it is not easy to specify a direct utility function that predicts 'Giffen behaviour' and simultaneously satisfies the basic axioms of consumer theory. With this problem in mind, we here adopt an alternative 'dual' approach, starting with an indirect utility function (IUF).

P.G. Moffatt (✉)
School of Economics, University of East Anglia, Norwich NR4 7TJ, United Kingdom
e-mail: p.moffatt@uea.ac.uk

W. Heijman and P. Mouche (eds.), *New Insights into the Theory of Giffen Goods*, Lecture Notes in Economics and Mathematical Systems 655,
DOI 10.1007/978-3-642-21777-7_10, 

The great advantage of using this dual approach is that it is possible for both the IUF and the related Marshallian demands to be expressible in explicit closed form.

## 2 The Indirect Utility Function

We assume that just two goods, good 1 (price $x$) and good 2 (price $y$), are available to the consumer. Since we are interested in price effects, we normalise the consumer's income to unity. Let $V(x, y)$ be the consumer's indirect utility function (IUF), i.e. the maximum utility attainable when the prices of the two goods are $x$ and $y$.

The IUF is assumed to satisfy the standard requirements of consumer theory, first, that $V(x, y)$ be monotonic decreasing in both price variables:

$$V_x \equiv \frac{\partial V}{\partial x} < 0, \quad V_y \equiv \frac{\partial V}{\partial y} < 0, \quad \text{in} \quad x > 0,\ y > 0\,; \tag{1}$$

and second, the requirement of 'quasi-convexity', i.e. that the contours of the IUF be convex to the origin. A clear intuitive proof that quasi-convexity is a necessary property of any IUF is provided by Suen [5].

## 3 Roy's Identity: A Geometric Interpretation

Roy's identity is the formula used to obtain Marshallian demand functions (i.e. utility maximising quantities expressed in terms of prices) directly from the IUF. Denoting the Marshallian demand functions by $g_1(x, y)$ and $g_2(x, y)$, Roy's identity is:

$$g_1(x, y) = \frac{V_x}{xV_x + yV_y}, \quad g_2(x, y) = \frac{V_y}{xV_x + yV_y}. \tag{2}$$

For good 1 (whose price is $x$) to be a *Giffen good*, $g_1(x, y)$ must be *increasing* in $x$ in some region, $\Gamma$ say, of the positive $(x, y)$ quadrant. The features required of the IUF for this to occur are most easily seen through the following geometric interpretation of Roy's identity.

Consider an IUF satisfying the asssumptions stated in Sect. 2. Let $P(x_1, y_1)$ be a point on any contour $V(x, y) = \text{cst}$. The gradient of the tangent at P is

$$dy/dx = -[V_x/V_y]_1, \tag{3}$$

where the suffix 1 indicates evaluation on at $x = x_1$, $y = y_1$. The equation of the tangent at P is then

$$y - y_1 = -[V_x/V_y]_1(x - x_1), \tag{4}$$

and this intersects the axis $y = 0$ at the point $(h, 0)$, where

$$h = x_1 + \frac{y_1}{[V_x/V_y]_1} = \left[\frac{xV_x + yV_y}{V_x}\right]_1. \tag{5}$$

Considering the equations (2), we see that the Marshallian demand for good 1 at prices $(x_1, y_1)$ is just the reciprocal of $h$:

$$g_1(x_1, y_1) = h^{-1}. \tag{6}$$

Hence we see that the Marshallian demand for good 1 can be interpreted as the reciprocal of the horizontal intercept of the tangent at P. Likewise, the Marshallian demand for good 2 is the reciprocal of the vertical intercept of the same tangent.

It follows that the condition required for giffenity of good 1, say, is that the horizontal intercept must move *to the left* when $x$ increases. That is, for giffenity, $\partial h/\partial x_1$ must be negative in some region $\Gamma$ of the $(x, y)$ quadrant. Another way of viewing this requirement is that the tangent gradient (3) must become more negative as $x$ increases, and must do so sufficiently rapidly for neighbouring tangents to intersect *above* the axis $y = 0$. This type of behaviour, involving local downward convergence of adjacent contours, is indicated schematically in Fig. 1.

# 4 Construction of an IUF Exhibiting Giffenity

## 4.1 Construction of a Family of Indifference Curves in Price-Space

We wish to construct a family of contours $V(x, y) = \text{cst.}$, having the 'downward convergence' property illustrated in Fig. 1. To this end, we follow Moffatt [1]

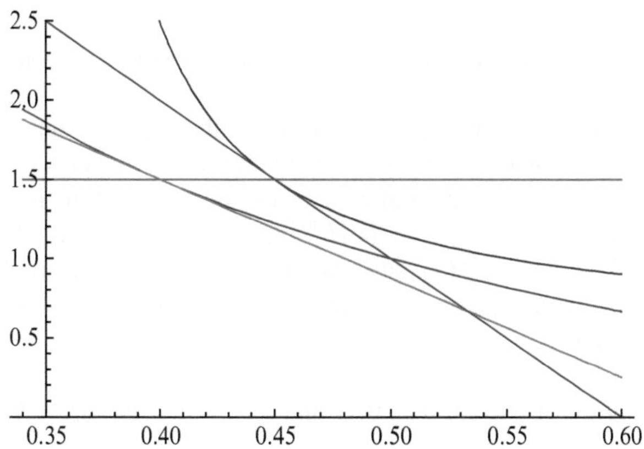


**Fig. 1** Schematic diagram: the tangents to two adjacent contours at the same value of $y$ intersect above the $y$-axis, a requirement for giffenity of good 1

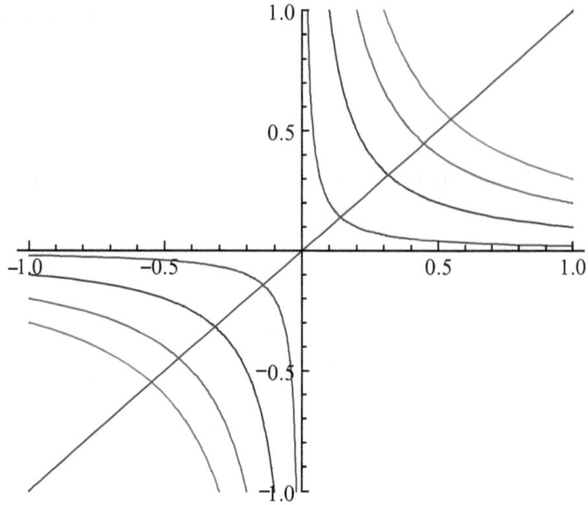


**Fig. 2** The family of hyperbolæ $xy = c^2$

in constructing a suitable family of hyperbolic contours exhibiting this property in some sub-region of the quadrant $(x > 0,\ y > 0)$. First consider the family of hyperbolæ $H : xy = c^2$ (Fig. 2).This has two branches:

$$H^+ : xy = c^2,\ x > 0,\ y > 0, \quad \text{and} \quad H^- : xy = c^2,\ x < 0,\ y < 0. \tag{7}$$

We shall be concerned with just the branch $H^+$ in what follows. With $y = c^2/x$, we have $y' \equiv dy/dx = -c^2/x^2$ and $y'' \equiv d^2y/dx^2 = 2c^2/x^3$. So at the point $(c, c)$ on $H^+$, $y' = 1$, $y'' = 2/c$ and the curvature at this point is

$$\kappa = \frac{y''}{(1 + y'^2)^{3/2}} = \frac{1}{\sqrt{2}\, c}. \tag{8}$$

Now consider the family of (upper-branch) hyperbolæ (Fig. 3):

$$H_a^+ : (x - a)(y - a) = c^2,\ \ x > a,\ y > a,\ \text{for variable}\, a > 0 \text{ and fixed } c. \tag{9}$$

As $a$ increases, $H_a^+$ simply moves out along the diagonal line $x = y$. Now however, we may achieve the required steepening of the tangents (as in Fig. 1) by choosing $c$ to be not constant, but rather a positive decreasing function of $a$, i.e. such that $c > 0, dc/da < 0$. The hyperbola $H_a^+$ intersects $y = x$ where $(x - a)^2 = c^2$, i.e. where $x = a + c(a) = f(a)$, say, and this function must clearly be monotonic increasing with $a$, in order to ensure that the curves of the family do not intersect; thus we require that

$$df/da = 1 + dc/da > 0, \quad \text{and so} \quad -1 < dc/da < 0. \tag{10}$$

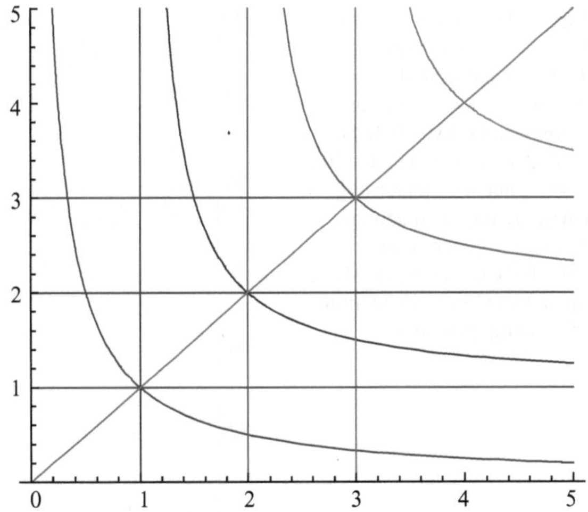


**Fig. 3** The curves $H_a^+$ (9) for $c = 1$ and $a = 0, 1, 2, 3$. As $a$ increases, the hyperbola is simply shifted out along the diagonal line $y = x$. The asymptotes of the four hyperbolæ are also shown

The simplest choice is

$$c(a) = 1 - \lambda a, \text{ where } 0 < \lambda < 1, \tag{11}$$

(to satisfy the inequalities (10)). Since $c(a) > 0$, we also require that $\lambda a < 1$, so this choice of $c(a)$ is limited to the range

$$0 < a < \lambda^{-1}. \tag{12}$$

Figure 4 shows a set of the curves $H_a^+$ with this choice of $c(a)$, which now indicate the downward convergence of the contours above the diagonal that is symptomatic of potential giffenity in good 1.

## 4.2 Further Considerations

Actually (11) is not ideal as a choice of $c(a)$, because it would require that we match smoothly to a different function to cover the region below the curve $xy = 1$. We can circumvent this by simply shifting the curves of Fig. 4 down the diagonal until this lowest member passes through the origin. This is achieved by the choice[1]

[1]The vanishing of $c$ for $a = (1 - \lambda)/\lambda$ means that the particular contour $a = (1 - \lambda)/\lambda$ has a right-angled corner at the point $(\mu, \mu)$, where $\mu = (1 - \lambda)/\lambda$. This singularity may be removed by adopting a modified $c(a)$ such that $c^2 = (1 - \lambda - \lambda a)^2 + \epsilon^2$, for suitably small $\epsilon$. We shall not pursue this refinement here.

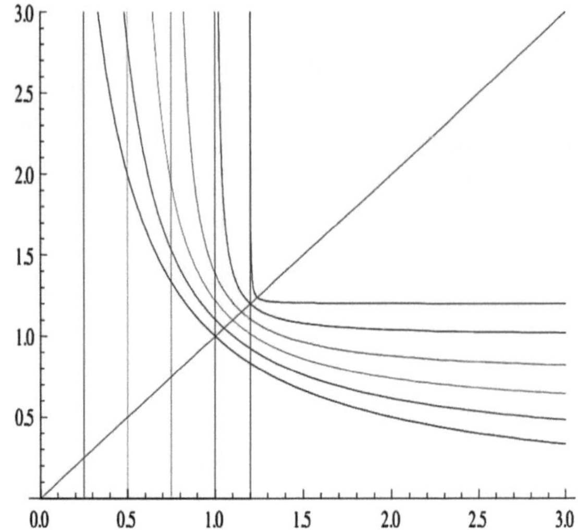


**Fig. 4** The curves $H_a^+$ (9) for $c = 1 - \lambda a$, with $\lambda = 4/5$ and $a = 0,\ 0.25,\ 0.5,\ 0.75,\ 1,\ 1.2$. Now, as $a$ increases, the hyperbola is shifted along the diagonal line $y = x$, but with increasing curvature on this diagonal. The convergence of the hyperbolæ in the region above the diagonal is an indication of potential giffenity

$$H_a^+ :\ (x-a)(y-a) = c^2, \text{for variable } a \geq -1, \text{ and } c(a) = 1 - \lambda - \lambda a, \quad (13)$$

with $0 < \lambda < 1$. Note that when $a = -1$, $c(-1) = 1$, and the corresponding hyperbola $(x+1)(y+1) = 1$ has 'centre' at $(-1,-1)$ and passes through $(0,0)$. Also, although the downward convergence of contours can be expected only for $c(a) > 0$, i.e. for $a < (1-\lambda)/\lambda$, we may in fact adopt the form (13) also for $a > (1-\lambda)/\lambda$, where the curvature on $x = y$ is then $-(\sqrt{2}\ c(a))^{-1}$ and *decreases* with increasing $a$. These properties are evident in Fig. 5, which shows this family of hyperbolæ, for the choice $\lambda = 4/5$. For this value, the point of maximum curvature is $(\mu, \mu)$, where $\mu \equiv (1-\lambda)/\lambda = 1/4$, so for $a < 1/4$, the curves are converging downwards (above the diagonal line $x = y$), and we may anticipate Giffen behaviour in a region above this diagonal near the point $(1/4, 1/4)$, where this downward convergence is strong. For $a > 1/4$, the contours show downward *divergence* above the diagonal, and giffenity cannot occur here.

Let $(X, Y)$ be a point on $H_a^+ :\ (x-a)(y-a) = c(a)^2$, so that

$$X = a + c(a)^2/(Y-a). \quad (14)$$

The gradient of $H_a^+$ at $(X, Y)$ is

$$\left[\frac{dy}{dx}\right]_{(X,Y)} = -\frac{c(a)^2}{(X-a)^2} = -\frac{(Y-a)^2}{c(a)^2}, \quad (15)$$

so the tangent to $H_a^+$ at $(X, Y)$ has equation

$$y - Y = -(Y-a)^2(x-X)/c(a)^2. \quad (16)$$

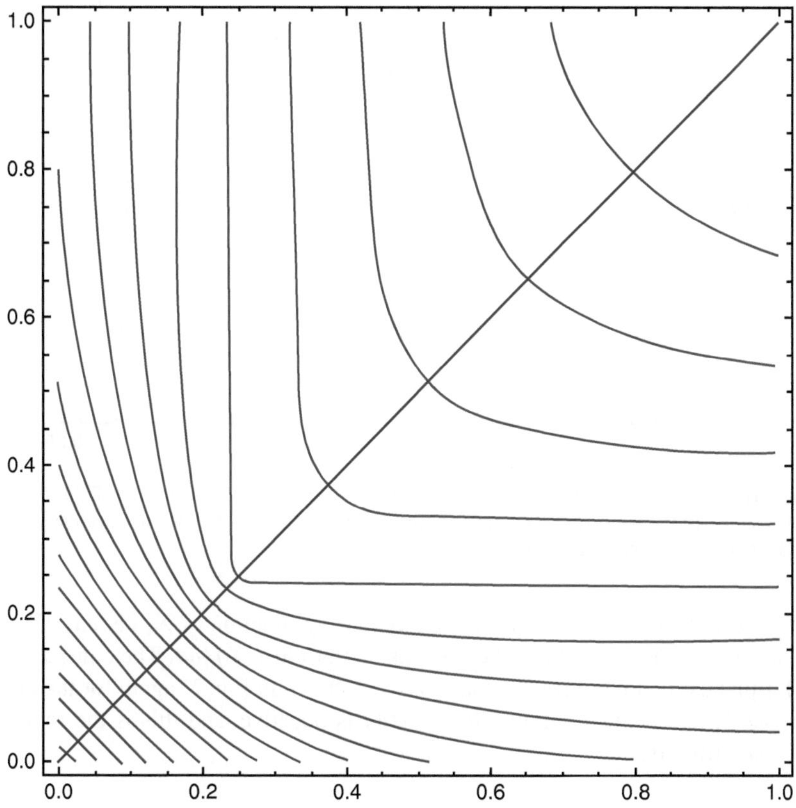


**Fig. 5** The family of curves $H_a^+$ defined by (13), for the choice $\lambda = 4/5$. The curve $a = 1/4$ (not shown in the figure) exhibits a corner at the point $(1/4, 1/4)$

This cuts the axis $y = 0$ at the point $(h, 0)$, where (cf. (5))

$$h = X + \frac{c(a)^2 Y}{(Y-a)^2} = a + \frac{c(a)^2}{Y-a} + \frac{c(a)^2 Y}{(Y-a)^2}, \tag{17}$$

which simplifies to

$$h(a, Y) = a + \frac{c(a)^2(2Y-a)}{(Y-a)^2}. \tag{18}$$

Figure 6 shows $h(a, Y)$ as a function of $a$ with $c(a) = 1 - \lambda - \lambda a$, for the choice $\lambda = 4/5$ and for seven values of $Y$ greater than $(1-\lambda)/\lambda (= 1/4)$. A range of negative gradient is apparent on the curves for which $0.25 < Y < 0.36$.

Now we are interested in the variation of $h$ with $X$ at constant $Y$, i.e. in

$$\partial h(a, Y)/\partial X = (\partial h(a, Y)/\partial a)(\partial a/\partial X). \tag{19}$$

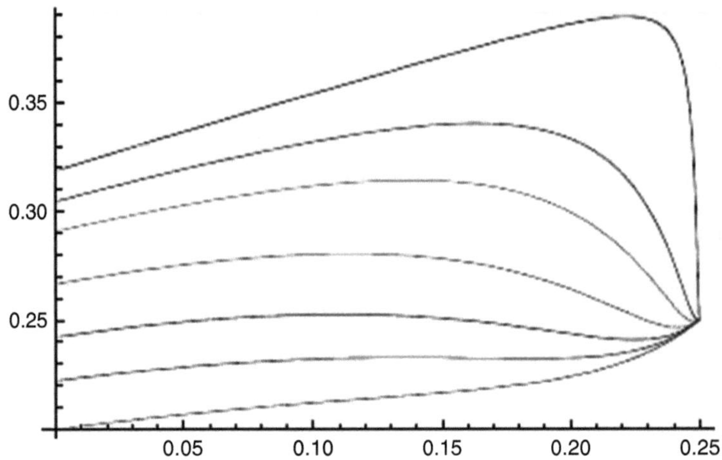


**Fig. 6** The function $h(a, Y)$ for $\lambda = 4/5$, and $Y = 0.251, 0.2625, 0.275, 0.3, 0.33, 0.36, 0.4$ (values increasing downwards). The uppermost five curves all show regions of negative gradient; this is the region of potential giffenity

It is obvious that, at constant $Y$, $a$ is an increasing function of $X$, i.e. $\partial a/\partial X > 0$. So the sign of $\partial h(a, Y)/\partial X$ is the same as that of $\partial h(a, Y)/\partial a$. The contour in the $(a, Y)$-plane on which $\partial h(a, Y)/\partial a$ is zero is shown in Fig. 7 (again for the choice $\lambda = 4/5$); inside this contour, $\partial h(a, Y)/\partial a$ is negative, and this is the region of potential giffenity.

## 4.3 The IUF in Explicit Form

Now we are in a position to obtain an explicit form for the IUF exhibiting giffenity. First note that for each $a$ we have a unique upper-branch hyperbola given by

$$H_a^+ : \ (x - a)(y - a) = (1 - \lambda - \lambda a)^2. \tag{20}$$

Thus, $a$ can be regarded as a label, constant on each $H_a^+$. Conversely, the member of the family $H_a^+$ passing through any given point $(x, y)$ is labeled by a root of (20) regarded as a quadratic equation for $a$. The relevant root (which vanishes as it should when $xy = (1 - \lambda)^2$) is

$$a = \frac{x + y - 2\lambda(1 - \lambda) - S(x, y)}{2(1 - \lambda^2)}, \tag{21}$$

where

$$S(x, y) = \sqrt{[x + y - 2\lambda(1 - \lambda)]^2 - 4(1 - \lambda^2)[xy - (1 - \lambda)^2]}. \tag{22}$$

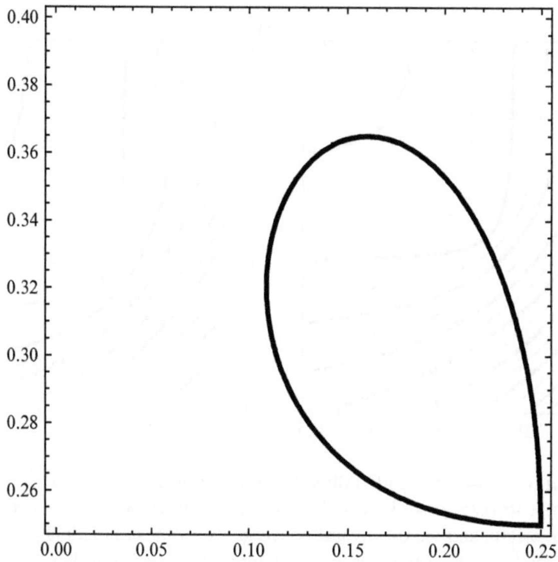


**Fig. 7** Contour $\partial h/\partial a = 0$ in the $(a, Y)$-plane ($\lambda = 4/5$, $0 < a < 1/4$, $1/4 < Y < 2/5$); $\partial h/\partial a$ is negative inside this critical contour

The other root (with $-S$ replaced by $+S$) corresponds to the lower branch $H_a^-$, and can be ignored. The contours $a(x, y) = \text{cst.}$ should be effectively the same as those in Fig. 5. Figure 8 show these contours for two values of $\lambda$ plotted directly from (21). Note how, as $\lambda$ increases towards unity, the point of maximum curvature $(\mu, \mu)$, where $\mu = (1 - \lambda)/\lambda$, moves towards the origin, and the downward 'giffenity-promoting' convergence of the contours above the diagonal $x = y$ increases; this trend persists as $\lambda$ increases further towards unity.

The function $a(x, y)$ is clearly an increasing function of $x$ and of $y$, i.e. $\partial a/\partial x > 0$, $\partial a/\partial y > 0$. The IUF must, of course, be decreasing in $x$ and $y$. Accordingly, let $V(a) = -a$, giving rise to the following IUF:

$$V(x, y) = -\frac{x + y - 2\lambda(1 - \lambda) - S(x, y)}{2(1 - \lambda^2)}, \tag{23}$$

where $S(x, y)$ is as defined in (22). Equation (23) is an IUF with the desired properties, i.e. being continuous and quasi-convex over the whole quadrant, satisfying the inequalities (1), and exhibiting the property of giffenity (in good 1) in the region indicated by Fig. 7.

The contours of this IUF $V(x, y)$ are shown in Fig. 9, again for the representative choice $\lambda = 4/5$; and, superposed on this, the closed contour (above the diagonal) within which the giffenity property is satisfied (obtained by substituting (21) in the

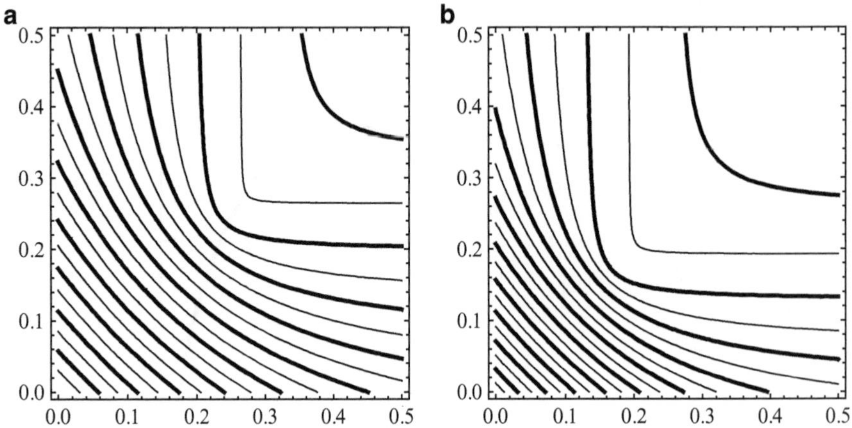


**Fig. 8** Contours $a(x, y) =$ cst., plotted directly from the function (21). (**a**) $\lambda = 0.8$, $\mu(= (1-\lambda)/\lambda) = 0.25$ ; (**b**) $\lambda = 0.85$, $\mu = 3/17 = 0.176\ldots$. Note how the critical ('corner') point $(\mu, \mu)$ moves towards the origin, and the downward convergence of the contours above the diagonal line $x = y$ increases (for $y < \mu$), with increase of $\lambda$, a trend that persists as $\lambda$ increases towards unity

**Fig. 9** Contours of the IUF given by (23) (with $\lambda = 4/5$) and (superposed) the region $\Gamma$ of giffenity of good 1 (above the diagonal); by symmetry, the region of giffenity of the good 2 is obtained, as shown, by simple reflection in the diagonal $x = y$

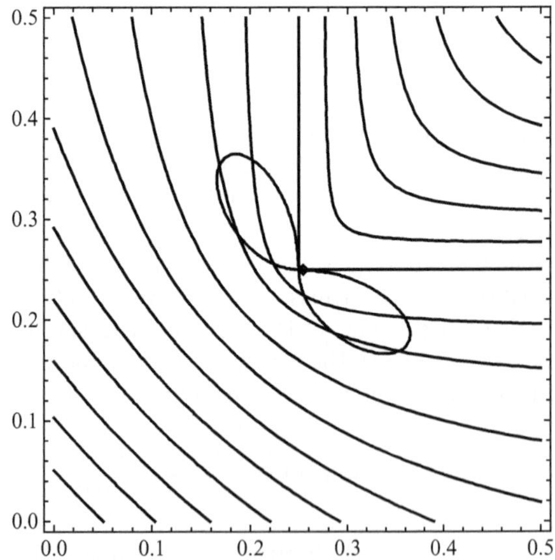


expression for $\partial h(a, Y)/\partial a$ obtained by differentiating (18)). This is the region $\Gamma$ referred to in Sect. 3. Note that, by the symmetry of the IUF about the diagonal, good 2 also exhibits giffenity in the similar region that is the mirror image of $\Gamma$ in the diagonal, as shown in the figure.

### 4.4 Marshallian Demands

Armed with an IUF in explicit form, we are now in a position to calculate the Marshallian demands via Roy's identity (2). The derivatives $V_x$ and $V_y$ can be obtained by straightforward differentiation of the above expression (23) for $V(x, y)$, and $g_1$ and $g_2$ may then be constructed from (2), and simplified. The Marshallian demand for good 1 so obtained is

$$g_1(x, y) = \frac{-x + y(1 - 2\lambda^2) + 2\lambda(1 - \lambda) + R(x, y)}{(x + y)[2\lambda(1 - \lambda) + R(x, y)] - 4xy\lambda^2 - (x - y)^2}, \tag{24}$$

where

$$R(x, y) = \sqrt{x^2 + y^2 + 4(\lambda - 1)[\lambda - 1 + \lambda(x + y)] + 2xy(2\lambda^2 - 1)}. \tag{25}$$

The Marshallian demand function (24) may be represented graphically in a variety of ways. First of all, a 3D plot (with $\lambda = 4/5$), is shown in Fig. 10. There is clearly a region near the centre of this plot in which the function appears to increase as $x$ increases – this is the "Giffen region". The position of this region may be more precisely identified my means of a contour plot. The contours $g_1(x, y) = \text{cst.}$ are shown in Fig. 11 (again with $\lambda = 4/5$), the values of $g_1$ on the various contours being as indicated; only the interesting region near the point $(1/4, 1/4)$ is shown, and the curve bounding the "Giffen region" within which $\partial g_1/\partial x$ is positive is superposed.

Finally, we may plot the Marshallian demand for good 1 against the price of good 1, for a fixed price of good 2. Such a plot is shown in Fig. 12, for which the price of good 2 has been set to $y = 0.30$. This plot clearly shows that when the price of good 1 is between 0.18 and 0.24, the function is positively sloped, confirming giffenity beyond any doubt.

In this section we have focused attention on the Marshallian demand for good 1. However, note that the symmetry of $V(x, y)$ implies that $g_2(x, y) \equiv g_1(y, x)$, so exactly the same patterns would be observed for good 2.

## 5 A Direct Utility Function Exhibiting Giffenity

It is interesting that the class of function developed in Sect. 4 for the IUF can also be used to construct a *direct* utility function that can predict giffenity. In this section, let us therefore redefine $x$ and $y$ as the quantities consumed of the two goods and consider the direct utility function $U(x, y)$.

Recall that in (21), we have a function that is increasing in $x$ and $y$, and it was this function to which we applied a sign reversal in order to obtain our IUF (23). Let us now use (21) itself as the direct utility function:

$$U(x, y) = \frac{x + y - 2\lambda(1 - \lambda) - S(x, y)}{2(1 - \lambda^2)}, \tag{26}$$

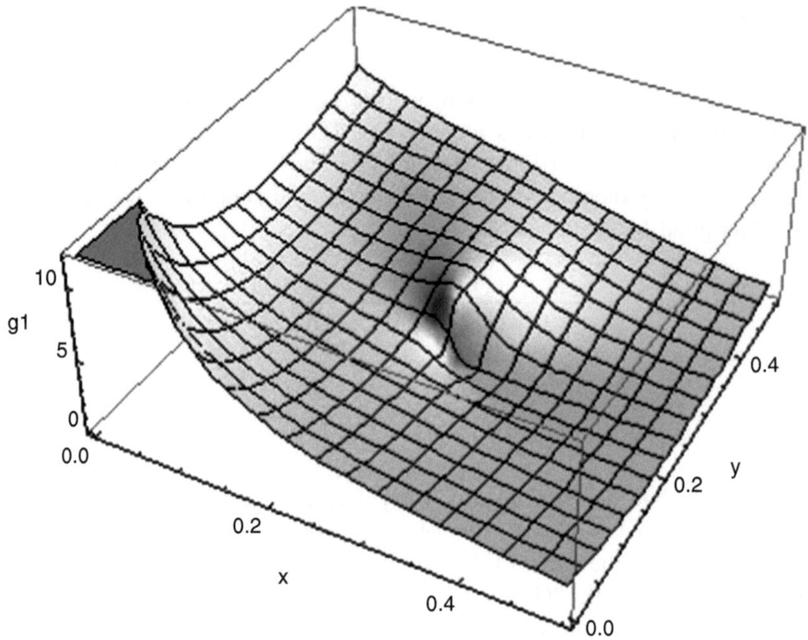


**Fig. 10** A 3D Plot of the Marshallian demand function $g_1(x, y)$, obtained from (23) and (2) with $\lambda = 4/5$

where $S(x, y)$ is still as defined in (22). Clearly, for any given $\lambda$, (25) has an indifference map which is identical to that of the IUF (23).

In Fig. 13, we plot the contours of (26) with $\lambda = 0.9$ (the corner singularity being now at $(1-\lambda)/\lambda = 1/9$), and two budget lines which show the effect of a rise in the price of good $x$. The upper budget line applies before, and the lower one after, the price rise. The important point is that the horizontal position of the point of tangency is further to the right following the price rise, implying that $x$ is a Giffen good at the assumed price combination. Since (26) is symmetric in $x$ and $y$, price combinations can also be chosen for which good $y$ is Giffen.

In [2], it is proved that the phenomenon demonstrated above is symptomatic of a general result: namely, that any two-good IUF that exhibits giffenity in one of the goods can be transformed, simply by reversing the sign, into a direct utility function that exhibits giffenity in the *other* good.

## 6 Conclusion

The purpose of this chapter has been to construct an indirect utility function (IUF) that predicts giffenity. One considerable advantage from this approach is that, given that such a function can be found, Roy's identity can be used to obtain the Marshallian demands explicitly.

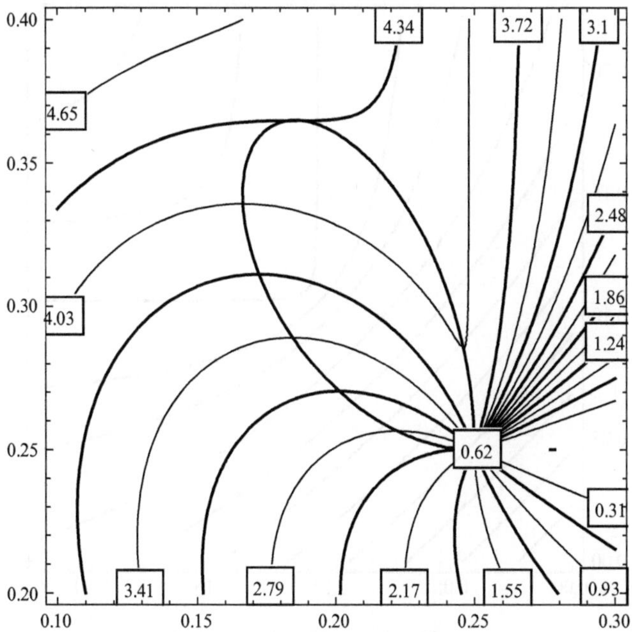


**Fig. 11** Contours of $g_1(x, y)$ in the region $(0.1 < x < 0.3, 0.2 < y < 0.4)$, obtained from (23) and (2) with $\lambda = 4/5$. The curve bounding the Giffen region $\Gamma$, within which $\partial g_1/\partial x > 0$, is superposed; note that the tangents to the contours are horizontal on this curve

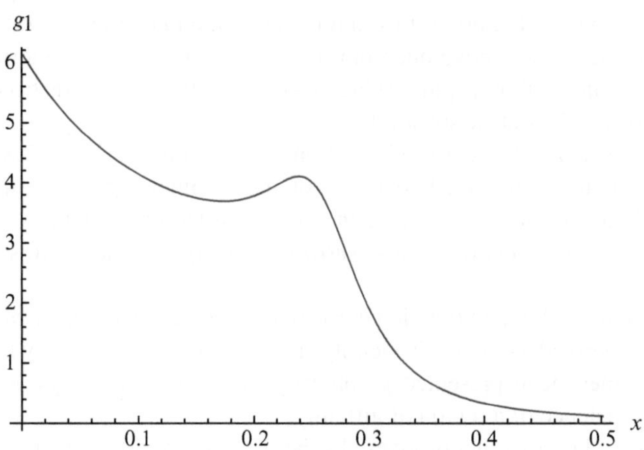


**Fig. 12** A plot of the the Marshallian demand function $g_1(x, y)$ against $x$, with $\lambda = 4/5$ and $y$ set to 0.30

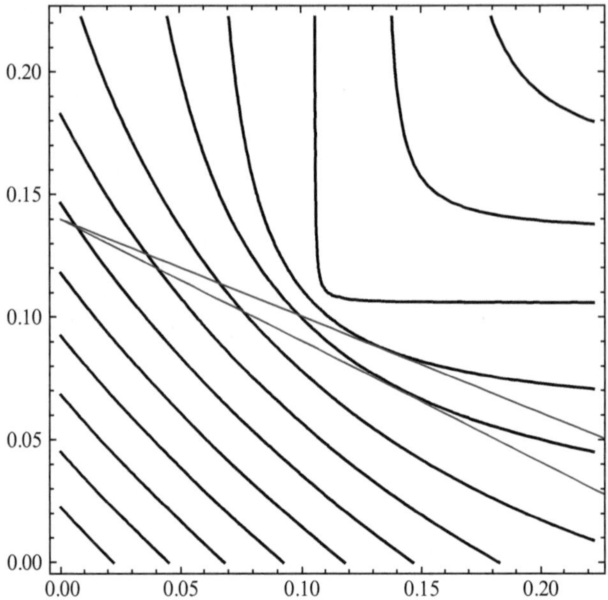


**Fig. 13** The direct utility function (26) with $\lambda = 0.9$. The two budget lines represent the positions before and after a rise in the price of good $x$. The lower budget line applies after the price rise. It is seen that the horizontal position of the point of tangency is further to the right following the price rise

A family of such IUFs has indeed been found, dependent on the parameter $\lambda$ $(0 < \lambda < 1)$. The key feature of this family is that the maximum curvature of each contour increases as we move out from the origin. An economic interpretation of this feature would be that as purchasing power rises, the two goods become closer complements, and less close substitutes.

An example from this class of function (for the choice $\lambda = 0.8$) has been presented graphically in Fig. 9. Because of the symmetry of the function, both of the goods are Giffen goods, and the sets of price combinations over which giffenity applies to each good are also shown in this figure. These two sets are non-overlapping.

The Marshallian demand function corresponding to the IUF represented in Fig. 9 has then been derived using Roy's identity, and presented algebraically in (24). This function has then been presented graphically in a variety of ways, all of which confirm the existence of a region of giffenity.

The chosen class of functions can also be used to provide Direct Utility Functions, and, as demonstrated in Fig. 13 (for the choice $\lambda = 0.9$), such an approach can also provide a demonstration of giffenity. However, it is not in general possible to derive explicit expressions for the Marshallian demands starting from such a Direct Utility Function. The fact that this is, by contrast, straightforward when starting from the IUF provides the rationale for the approach adopted in this chapter.

## References


1. Moffatt P. G. 2002. Is Giffen behaviour compatible with the axioms of consumer theory? *Journal of Mathematical Economics* 37:259–267.
2. Moffatt, P.G., Moffatt, H.K. 2010. Giffen Goods: a Duality Theorem. Mimeo, University of East Anglia.
3. Sörensen P. 2006. Simple Utility Functions with Giffen Demand *Economic Theory* 31:367–70.
4. Spiegel U. 1994. The Case of a Giffen Good. *Journal of Economic Education* 25:137–147.
5. Suen W. 1992. A diagrammatic proof that indirect utility functions are quasi-convex *Journal of Economic Education* 23:53–55.
6. Weber C.E. 1997. The case of a Giffen Good: Comment *Journal of Economic Education* 28:36–44.
7. Weber C.E. 2001. A Production Function with an Inferior Input *The Manchester School* 69:616–622.
8. Wold H., Jureen L. 1953. *Demand Analysis* New York: Wiley.

# Close Substitutes and Upward-Sloping Demand Curves

**Jan Rouwendal**

**Abstract** It is well known that the 'law of demand' may fail in the case of inferior goods. It seems less appreciated that we can also have upward-sloping demand curves in other circumstances. This paper looks at cases in which demand refers to categories of goods that are close substitutes. Three situations are considered. The first concerns a quality differentiated commodity of which one variety is consumed. Examples of such products abound. The second situation refers to the use of a durable of which two units are present in the household. Cars are a natural example. The third concerns the long run demand curve for the use of a durable that is available in various qualities. One may again think of the car, but there are many other possibilities. In each of these cases economic theory does not exclude the possibility that demand functions are upward-sloping. Concrete specifications where this happens are provided.

## 1 Introduction

It is well known that the "law of demand"[1] can only be derived from economic theory for the case of normal goods. For inferior goods the substitution effect and the income effect of a price change have opposite signs, and this makes it impossible to exclude upward-sloping demand curves. Marshall famously attributed the empirical

[1] I take this law here to state that the partial derivative of a demand function with respect to the own price is always negative.

J. Rouwendal (✉)
Department of Economics, Vrije Universiteit, De Boelelaan 1105, 1081HV Amsterdam, The Netherlands
e-mail: jrouwendal@feweb.vu.nl

W. Heijman and P. Mouche (eds.), *New Insights into the Theory of Giffen Goods*, Lecture Notes in Economics and Mathematical Systems 655,
DOI 10.1007/978-3-642-21777-7_11, © Springer-Verlag Berlin Heidelberg 2012

evidence for such anomalous demand curves to Giffen. More recently, such evidence was provided by [5].

Giffen goods occur in the pure theory of consumer behavior where consumers spend their given budget on nondurable[2] commodities that are completely homogeneous. If Giffen goods exist under such circumstances, one may conjecture that they may show up even easier under less pure, and perhaps more realistic circumstances in which demand does not refer to a single completely homogeneous commodity, but to a group of commodities that are close substitutes. This is the conventional situation in applied demand analysis. Empirical studies of consumer demand systems distinguish a number of commodity groups that is small in comparison to the enormous amounts of products that are available in modern supermarkets. These commodity groups are treated as the completely homogeneous goods whose demand is studied in microeconomic theory. See, for instance, [1] for an empirical demand system covering all nondurable consumption, and [6] for studies that focuses on food. In both studies 22 categories are distinguished, which is large compared to many other empirical studies. Nevertheless, it is clear that there is a large amount of product heterogeneity present even in the most homogeneous category of food products considered by LaFrance. In both studies all commodity groups are treated as a single homogeneous commodity, although it is clear that there exist abundant possibilities for substitution within groups. In reaction to a price change, consumers may substitute towards a lower quality version of the commodity, and it may well be the case that such substitution mitigates the effect of the price change on the quantity demanded. For instance, an increase in the price of all kinds of potatoes may well result in a shift of demand towards the cheaper varieties, while the quantity demanded does not change that much because people need approximately the same amount of calories. Intuitively, one expects still a net decrease in the total demand for potatoes, but the question we will consider in this paper is what economic theory has to say about this net effect.

Another situation that occurs frequently is that the total demand for a particular homogeneous commodity derives from different demand functions. One may think of the demand for energy that is used for cooking as well as for heating the home. The two may to some extent be substitutes as cooking contributes to heating and eating outside decreases the need for heating the home. An alternative example is the demand for kilometers in a household with multiple cars. In reaction to a changing gasoline price, households may substitute the use of one car for the other and although one expects intuitively that the net result is a decrease in household demand for gasoline, it remains to be seen if economic theory predicts this effect.

---

[2]Later in this paper we will introduce durable commodities whose use can be studied as a continuous variable that is available at a constant unit price, but also require the payment of a fixed cost, which is independent of the use of the durable. It is conceivable that the price of such a commodity changes. In the case of price increases this may lead to increasing demand curves. Housing seems to be a relevant example. However, these intertemporal issues will not be studied here.

A third situation of potential interest is one in which the consumer can chose the quality of a particular durable, while this quality is also related to the intensity of fuel use. One may think about an air conditioning or heating system, or, again, about the private car. See, for instance, [4] or [3] for early studies. In reaction to a change in the fuel price, the consumer may decide to buy a lower quality of the durable that uses less fuel. Due to this effect, one expects the long run demand for the fuel used by this durable to be somewhat less elastic than the short run demand. Again, it remains to be seen if this effect is predicted by economic theory.

It is the purpose of this paper to study the possibility that substitution within groups of goods results in upward-sloping demand curves. I will consider situations in which consumers can choose among two or more varieties of goods that are close substitutes and consider the effect of a change in a price shifter that affects the prices of all goods in the group. The increase in the prices of all commodities in the group induces substitution behavior that may well mitigate the quantity effects of the price change. In the sections that follow we consider three different models in which there are closely related consumption goods that may reasonably be considered as a homogeneous group. These models represent the three situations just discussed.

The first model concerns a quality-differentiated non durable consumption good. Quality is treated as a continuous scalar variable, and the consumer chooses one quality. The unit price of the commodity depends on the quality, and on a price shifter. Conditional upon the choice of quality, this is the standard textbook model of consumer theory. When quality can be chosen this becomes a non-standard situation which is arguably more realistic for many differentiated commodities that are presented on the shelves of our supermarkets than the homogeneous commodity of undergraduate microeconomic theory. In the context of that model, the total quantity demanded is studied as a function of the price shifter. The price shifter may be interpreted as the price of an input that is used in all varieties. For instance, if the model refers to strawberry jam, the price shifter may be the price of fresh strawberries in the summer of 2009, which may have been particularly low because of the good weather conditions in that season.

The second case we consider is two durables using the same fuel. The natural example to think of is a multiple car household with two gasoline cars. What happens when the fuel price changes? I formulate a standard microeconomic model for this situation and derive an expression for the net effect of the increase in the fuel price on the quantities demanded for both goods, as well as for the total amount of gasoline demanded by the consumer. The sign of the effect of a change in the fuel price on total demand turns out to be ambiguous.

In the third example, I consider a consumer who has to choose the quality of a durable that uses a particular kind of fuel. The focus is on the long run demand for the fuel. More specifically, the question of interest is how the demand for fuel is affected by changes in its price when the quality of the durable can be adjusted. Again, the private car may be the example that comes to mind first. The model implies an expression for the long run demand for fuel that strongly suggests that it may be upward-sloping.

For all three cases concrete specifications will be provided that illustrate the existence of upward-sloping demand curves for particular ranges of parameter values. The point of these further elaborations is not that they represent reality, but rather that economic theory does not exclude their possible existence. The 'law of demand' may fail in circumstances in which conventional Giffen behavior is absent. To underline this point, the concrete examples are shown to work with quasilinear preferences, where income effects are absent and the conventional type of Giffen behavior is therefore excluded.

Before concluding this introductory section it is important to note, in the context of this book, that in this chapter not all the underlying mathematical assumptions will be made explicit. For instance, when demand functions are derived it is just assumed that these are single valued, and when an identity is differentiated it is just assumed that the functions involved are differentiable.

# 2 A Model of Quality Choice

## 2.1 A General Model

In this section we consider a simple model of quality choice. The consumer has preferences that can be described by a utility function $u$ with three arguments: the quantity consumed of a composite commodity $c$, the quantity consumed of the commodity of interest $q$, and the quality $k$ of the consumed units of the commodity of interest.

$$u = u(c, q, k). \tag{1}$$

The utility function is assumed to be well-behaved. Well-behavedness will not be rigorously defined but implies certainly that the utility function is twice differentiable, quasi concave in $c$ and $q$ and increasing in $c$, $q$, and $k$. The budget constraint is:

$$c + p(\pi, k)q = y, \tag{2}$$

where $p$ denotes the unit price of the commodity of interest, which depends on a shift parameter $\pi$ and the quality chosen, and $y$ is total consumption expenditure.

Conditional on the choice of $k$ this is the standard utility function of the textbook two-good model. Quality may be vector valued, but in what follows we focus on the case in which it is a scalar.

The first-order conditions for utility maximization are:

$$\frac{\partial u}{\partial c} = \lambda, \tag{3}$$

$$\frac{\partial u}{\partial q} = \lambda p(\pi, k), \tag{4}$$

$$\frac{\partial u}{\partial k} = \lambda q \frac{\partial p}{\partial k}. \tag{5}$$

In these equations $\lambda$ denotes the marginal utility of income. The first two conditions are familiar. The third one says that the marginal utility of quality is proportional to the quantity consumed of the commodity of interest.

Solving the first order conditions leads to demand equations for the composite good and the quality and quantity of the commodity of interest:

$$c = c(\pi, y) \tag{6}$$

$$q = q(\pi, y) \tag{7}$$

$$k = k(\pi, y). \tag{8}$$

The parameter $\pi$ may reflect a tax levied on all variants of the heterogeneous commodity. A change in the tax would change the consumer prices of all varieties. Alternatively, it may be the price of an input used by all varieties. A change in the price of this input will be passed on to the consumers of the commodity.

We will now analyze the effect of a change in $\pi$ on the quantity demanded of the commodity of interest, and focus attention on the possible existence of Giffen behavior: an increase in the quantity demanded in response to an increase in the shift parameter. To do this, we first define the conditional demand for the commodity of interest as the demand function that results from taking the quality $k$ as given. In that situation, first order condition (5) is not relevant, and we can solve the system (2), (3) and (4) in the conventional (textbook) way. The conditional demand equation for the commodity of interest is:

$$q = q^c(p(\pi, k), y; k), \tag{9}$$

which differs from the (unconditional) demand equation (7) by having $k$ as an argument. The unconditional demand equation can be derived from (9) by substitution of the demand equation for quality (8):

$$q(\pi, y) = q^c(p(\pi, k(\pi, y)), y; k(\pi, y)). \tag{10}$$

To analyze the effect of a change in the shift parameter on $q$, differentiate the right-hand side of (10) with respect to $\pi$:

$$\frac{\partial q}{\partial \pi} = \frac{\partial q^c}{\partial p}\left(\frac{\partial p}{\partial \pi} + \frac{\partial p}{\partial k}\frac{\partial k}{\partial \pi}\right) + \frac{\partial q^c}{\partial k}\frac{\partial k}{\partial \pi}. \tag{11}$$

The question is: can the expression on the right-hand side be positive? Two possibilities exist, even if we take $\partial q^c/\partial p < 0$, which rules out the conventional Giffen goods, and $\partial p/\partial \pi$, $\partial p/\partial k > 0$. The first requires $\partial k/\partial \pi$ to be negative, the second that $\partial q^c/\partial k$ and $\partial k/\partial \pi$ have opposite signs. In both cases there is a positive term on the right-hand side of (11) that makes its sign indeterminate.

The model derived here can be related to any standard "textbook" model of consumer behavior by considering this standard model as conditional on the quality of the commodity $k$. The model can then be generalized by *(i)* further specifying one or more parameters of the utility function as a function of quality $k$ in such a way that utility is increasing in $k$ and indifference curves are convex, and *(ii)* specifying the price $p$ as an increasing function of quality $k$ and a price shifter $\pi$. The standard model that is the starting point must of course satisfy the conditions of demand analysis and care must be taken that optimal quality choice results in an interior solution. The next subsection provides a specific example.

The generalized model, with variable quality, is in fact closely related to the "hedonic" model of [8]. The only difference is that [8] assumed that only one unit of the quality differentiated commodity is consumed (as is natural when one studies, for instance, housing consumption), whereas I do not impose that restriction.

The analysis given above has resulted in the conclusion that the theory does not guarantee that the number of units consumed of the quality differentiated good decreases when the prices of all varieties of that good go up. In other words, if the price of all kinds of strawberries goes up because of bad weather conditions, economic theory as embodied in the present model does not imply that the demand for strawberries of a rational actor goes down. The reason is that the model developed here allows for substitution between different qualities of strawberries.

It is, of course, one thing to show that the sign of a partial derivative is indeterminate, and another to show a concrete example in which demand functions can be upward-sloping. Marshall's finding that the slope of ordinary (Marshallian) demand functions was left indeterminate by economic theory gained significance by his discussion of the empirical relevance of this possibility. However, the real test was a demonstration that upward-sloping demand functions could really be derived from the theory by means of a specific example. This test has also to be passed by the model developed here.

## 2.2 A Specific Example

To show that the model developed in the previous subsection may really imply upward-sloping demand functions, a simple example that demonstrates this will now be developed. We start with a demand equation for a quality-differentiated good that is linear in price and quality:

$$q = \alpha - \beta p(k) + \gamma k, \tag{12}$$

where the Greek letters denote positive coefficients. Income does not appear in this demand equation. Conditional on quality, the demand function is linear in the price and independent of income. The underlying textbook model results when $k$

is interpreted as a constant, and it can be shown to be consistent with economic theory.[3]

Note that demand $q$ is independent of income. The absence of an income effect rules out the possibility of the conventional Giffen goods, which can only exist for inferior goods. It implies also that the indirect utility function is the sum of income and Marshallian consumer surplus. Utility maximization with respect to quality choice is therefore equivalent to maximization of consumer surplus with respect to quality choice.

To derive consumer surplus, first compute the reservation price $p^{res}$, at which demand becomes equal to 0, as:

$$p^{res} = \frac{\alpha + \gamma k}{\beta}. \tag{13}$$

Consumer surplus $CS$ at an arbitrary price $p \leq p^{res}$ can now be determined as:

$$CS = \frac{\alpha^2 + 2\alpha\gamma k + \gamma^2 k^2 + \beta^2 p^2}{2\beta} - (\alpha + \gamma k)p. \tag{14}$$

The effect of a marginal change in the quality on consumer surplus is:

$$\frac{\partial CS}{\partial k} = \gamma \frac{\alpha - \beta p(k) + \gamma k}{\beta}, \tag{15}$$

which is positive. This implies that the consumer will choose the highest possible quality. However, there is a counteracting force when the unit price also depends on quality, as we will assume. To bring this out clearly, the dependence of price $p$ on quality $k$ has been made explicit in (15).

Economic theory does not imply much about the functional form of $p(k)$. The price must be increasing in quality, for otherwise every actor would opt for the highest quality and we would not observe product differentiation. For the specification of demand used here, the first order condition for optimal quality choice says that $\partial CS/\partial k$ must be zero, whereas the second order condition requires that $\partial^2 CS/\partial^2 k$ must be negative. One specification that allows for this possibility is the following quadratic one:

$$p(k) = \pi(\rho_0 + \rho_1 k + \rho_{kk} k^2). \tag{16}$$

In this equation $\pi$ is a shift parameter that reflects general cost conditions, for instance the real wage of labor involved in the production of the quality

---

[3]This can be done by verifying the integrability conditions for the implied complete two-equation demand system, or by checking the conditions for the indirect utility function (discussed below) to be consistent with economic theory. For discussion of these conditions, see for example [2] or [7] and the references cited there.

differentiated commodity. The coefficients $\rho$ are all positive. The expression in parentheses reflects the increasing amount of inputs (for instance, labor) needed to realize a variant with a higher quality.

The consumer maximizes her $CS$ by choosing the quality as well as the quantity of the commodity, while taking into account the price function (16). The first order condition is:

$$\frac{\partial CS}{\partial k} = q\frac{\partial p}{\partial k}, \tag{17}$$

which is (5) for the specification of the preferences considered here. Solving this condition for the optimal $k$ gives:

$$k = \frac{\gamma - 2\beta\rho_k\pi}{4\beta\rho_{kk}\pi}. \tag{18}$$

It is easy to verify that $\partial CS/\partial k$ is decreasing in $k$, while $q(\partial p/\partial k)$ is increasing in $k$. This implies that the second order condition for a maximum is satisfied. To further check the plausibility of the implied quality choice behavior, we investigate the effect of a higher shift parameter on the optimal quality choice. Using (18) we compute:

$$\frac{k}{\pi} = -\frac{\gamma}{4\beta\rho_{kk}\pi^2} \tag{19}$$

which implies that a higher value of $\pi$ implies a lower optimal quality.

Substitution of optimal quality choice (6)–(8) in the conditional demand equation (12) and in the price equation (16), leads to the unconditional demand equation $q(\pi, y)$ (compare (7)) for this specification of the model:

$$q = \frac{3\gamma^2 - 8\beta\gamma\rho_k + 4\beta\pi\left(4\alpha\rho_k k + \beta\pi(\rho_k^2 - 4\rho_0\rho_{kk})\right)}{16\beta\rho_k k\pi}. \tag{20}$$

This brings us, finally, in the position in which we can consider the crucial question whether in this apparently reasonable model the quantity demanded of the heterogeneous good can increase in response to an increase in the shift parameter $\pi$. For this purpose we compute the first derivative of demand with respect to $\pi$, and find:

$$\frac{\partial q}{\partial \pi} = -\beta\rho_0 + \frac{4\beta^2\rho_k^2\pi^2 - 3\gamma^2}{16\pi^2\beta\rho_{kk}}. \tag{21}$$

If the right-hand side of this equation is positive, the desired result has been shown. The right hand side can only be positive if the second term is positive and larger (in absolute value) than the first. This requires that the numerator of the second expression on the right-hand side is positive. We can choose the parameter values so that this condition is fulfilled. If this is the case, the whole expression on the right hand side becomes positive if $\rho_{kk}$ is small enough. Since I am free to do so, I have constructed an example in which the demand function for the heterogeneous good is upward-sloping as a result of substitution between varieties of the good.

Summarizing, it has been shown that a generalization of the textbook two good model to a situation in which one of the goods is available in a continuum of varieties, as in Rosen's (1974) model but without the restriction that only one unit of the heterogeneous good can be consumed, the number of units demanded can increase in response to an increase in the price of all varieties of the good.

# 3 Substitutes Using the Same Input

## 3.1 A Model of Using Two Durables That Are Substitutes

In this section we consider a different situation. Consider a household owning two cars. These two cars are substitutes for each other, in the sense that at least for some trips household members can choose between using one or the other car. The price of using a car is its variable cost which is determined by the fuel price and the fuel efficiency of the car. For simplicity we abstract from other determinants of variable cost. Moreover, we assume that both cars use the same fuel type, say gasoline. If the fuel price goes up, this affects the demand for kilometers driven by both cars in two ways: the own price goes up, which has a negative effect on the number of kilometers unless this commodity is a Giffen good, and the price of the substitute goes up, which has a positive effect. The two effects are present for the number of kilometers driven by both cars. The question we focus on in this section is: can the total effect of an increase in the gasoline price on the number of kilometers driven by the household be positive if the demands for kilometers of the individual cars are non-Giffen goods? Intuitively one expects the answer to be negative, but we have to see whether a formal analysis supports that conjecture.

To model this situation we assume again a utility function with three arguments: the quantity consumed of a composite commodity, $c$, and the kilometers driven by the two cars, $q_1$ and $q_2$:

$$u = u(c, q_1, q_2). \tag{22}$$

The budget constraint is:

$$c + p_1 q_1 + p_2 q_2 = y, \tag{23}$$

where the $p$'s denote the variable costs of the two cars and $y$ is the household's budget for non durable consumption. The variable costs are defined as:

$$p_i = \pi_i, \qquad i = 1, 2, \tag{24}$$

where $\pi$ is the gasoline price and $f$ is fuel use per kilometer, or fuel intensity. Solving the utility maximization problem in the conventional way leads to demand equations:

$$c = c(p_1, p_2, y), \tag{25}$$

$$q_1 = q_1(p_1, p_2, y), \tag{26}$$

$$q_2 = q_2(p_2, p_1, y). \tag{27}$$

The total demand for kilometers by this household, $q$, is the sum of the demands for kilometers by the two cars:

$$q(p_1, p_2, y) = q_1(p_1, p_2, y) + q_2(p_2, p_1, y), \tag{28}$$

and we are interested in the impact of an increase in the gasoline price on this demand:

$$\frac{\partial q}{\partial \pi} = \frac{\partial q_1}{\partial p_1} f_1 + \frac{\partial q_1}{\partial p_2} f_2 + \frac{\partial q_2}{\partial p_2} f_2 + \frac{\partial q_2}{\partial p_1} f_1. \tag{29}$$

If the demands for kilometers by the individual cars are not Giffen goods we have two negative terms on the right-hand side of this equation. Assuming that demand for kilometers by the two cars is normal and that the two cars are substitutes (in the Hicksian sense), implies that the two cross-derivatives are positive. The sign of the total effect is therefore indeterminate, and may be positive.

The intuitive reason for expecting the net effect on the total number of kilometers to be negative is that one expects the negative effect of an increase in the variable cost of car 1 on the demand of kilometers by that car to exceed, or at least be equal to, the positive cross effect on the demand for kilometers by car 2. This expectation is closely related to the concept 'substitution.' This intuition implies that, if one rewrites (29) as:

$$\frac{\partial q}{\partial \pi} = \left(\frac{\partial q_1}{\partial p_1} + \frac{\partial q_2}{\partial p_1}\right) f_1 + \left(\frac{\partial q_2}{\partial p_2} + \frac{\partial q_1}{\partial p_2}\right) f_2 \tag{30}$$

one expects both expressions in brackets to be negative. The question now is if this is implied by economic theory? The answer is negative, as will be shown by the example that follows.

### 3.2 An Example: Demand for Car Kilometers in Multiple Car Households

In this section I illustrate the potential for Giffen goods in the model developed in the previous section by specifying what is perhaps the most simple variant that incorporates the relevant issues. Consider a household owning two cars. In the short run the quality of these cars is taken as given. Preferences of a household owning two cars are described by the following indirect utility function:

$$v = g(y - F_1 - F_2) + \frac{1}{2}\frac{\beta_2(\alpha_1 - p_1)^2 + \beta_1(\alpha_2 - p_2)^2 - 2\gamma(\alpha_1 - p_1)(\alpha_2 - p_2)}{\beta_1\beta_2 - \gamma^2} \tag{31}$$

with $g(.)$ a concave function.[4] In this equation $F_1$ denotes the fixed cost of car $i$. In the special case in which $g$ is linear, preferences are quasilinear, which implies that demand for kilometers does not depend on income. Application of Roy's identity results in the following short run demand functions for car kilometres:

$$q_1 = \frac{(\alpha_1\beta_2 - \alpha_2\gamma) - \beta_2 p_1 + \gamma p_2}{g\prime(y - F_1 - F_2)(\beta_1\beta_2 - \gamma^2)}, \tag{32}$$

$$q_2 = \frac{(\alpha_2\beta_1 - \alpha_1\gamma) - \beta_1 p_2 + \gamma p_1}{g\prime(y - F_1 - F_2)(\beta_1\beta_2 - \gamma^2)}. \tag{33}$$

with $g\prime(.)$ the first derivative of $g$. These demand functions are linear in the two prices (variable costs).

It is clear from (32)–(33) that the substitution between kilometres driven by the two cars is indicated by $\gamma$. The second order condition for a constrained utility maximum is satisfied when $\beta_1$ and $\beta_2$ are positive and $\beta_1\beta_2 - \gamma > 0$. We do, of course, expect that the intercepts of both demand equations are positive, and to ensure this, we assume:

$$\alpha_1\beta_2 - \alpha_2\gamma > 0 \tag{34}$$

$$\alpha_2\beta_1 - \alpha_1\gamma > 0 \tag{35}$$

$$\beta_1\beta_2 - \gamma^2 > 0. \tag{36}$$

Note that all three conditions state – in different ways – that the substitution parameter should not be 'too large' for the model to make sense. One may interpret this as saying that two cars will only be used if substitution is not too easy.

The conditions (34)–(36) are necessary for demand for kilometres driven by both cars to be positive at prices that are sufficiently close to 0, but not sufficient. In particular, the second order condition for a maximum (the inequality (36)) can be satisfied with a negative intercept for one demand equation. And even with positive intercepts, the demand for car kilometres is (of course) not guaranteed to be positive at all prices.

The special case of *symmetric preferences*, in which $\alpha_1 = \alpha_2$ and $\beta_1 = \beta_2$ is a useful benchmark. In this case all three conditions (34)–(35) say that $\gamma < \beta_1 = \beta_2$. In non-symmetric cases $\gamma$ can be larger than one $\beta$, but it has to be smaller than the other. In that situation, the car with the largest $\beta$ must have the smallest $\alpha$.

---

[4]It is not difficult to verify that this function satisfies the requirements for an indirect utility function. The system of demand functions (25)–(27) is therefore also consistent with economic theory. Again, see for example, [2] and [7] and the references cited there.

The special case of interest arises when car 1 is *absolutely preferred* to car 2, that is: $\alpha_1 > \alpha_2$ and $\beta_1 \leq \beta_2$. In this case the first kilometre driven by car one contributes more to utility than the first kilometre driven by car 2, and the contribution to utility of additional kilometres driven decreases as much for car 2 as for car 1. As we just noted, in this case it is possible that $\gamma$ is larger than $\beta_2$. Now consider the situation in which the two cars are identical. Then we have $p_1 = p_2 = p$. In this situation an identical change in both variable costs, due to, for instance, a change in the price of gasoline results in an increase in the number of kilometres driven by car 2, as is easy to verify from (32)–(33). In other words: due to the fact that the substitution effect is larger than the own price effect, the net effect of an increase in the gasoline price on the demand for kilometers in car 2 is *positive*.

This effect is already counterintuitive, but does not yet establish the point we want to make here, namely that total demand for gasoline may increase when its price increases. To consider this issue at a somewhat more general level, return to the general model and add up the two demand equations (32) -(33). The result is:

$$q_1 + q_2 = \frac{\alpha_1\beta_2 + \alpha_2\beta_1 - \gamma(\alpha_1 + \alpha_2) + (\gamma - \beta_2)p_1 + (\gamma - \beta_1)p_2}{g\prime(\gamma - F_1 - F_2)(\beta_1\beta_2 - \gamma^2)}. \tag{37}$$

Using (24), we derive the effect of an increase in the prices of gasoline $\pi$ on total gasoline demand:

$$\frac{\partial(q_1 + q_2)}{\partial\pi} = \frac{(\gamma - \beta_2)f_1 + (\gamma_\beta)f_2}{g\prime(y - F_1 - F_2)(\beta_1\beta_2 - \gamma^2)}. \tag{38}$$

To make the possibility of a positive sign of the right-hand side of (38) as concrete as possible, fix the values of the left-hand sides of the inequalities (34) and (35) at arbitrary positive values, say $x_1$ and $x_2$. Solving the resulting equalities for $\alpha_1$ and $\alpha_2$ then gives:

$$\alpha_1 = \frac{\beta_1 x_1 + \gamma x_2}{\beta_1\beta_2 - \gamma^2} \tag{39}$$

$$\alpha_2 = \frac{\beta_2 x_2 + \gamma x_1}{\beta_1\beta_2 - \gamma^2}. \tag{40}$$

Since the left-hand sides of these equations are positive for arbitrary positive values for the $\beta$'s and $\gamma$, the inequalities (34) and (35) do not restrict our choices of these three parameters.

Now assume that car 1 is absolutely preferred to car 2, and that $\gamma - \beta_1$ is positive. We have to take care that the inequality (36) remains satisfied. Denoting $\beta_2$ as $\beta_2 = \gamma(1 + \epsilon)$, for some $\epsilon > 0$ it follows that we must have: $\epsilon > (\gamma - \beta_1)/\beta_1$. We can now rewrite the numerator of the right-hand side of (38) as: $-\epsilon f_1 + (\gamma - \beta_2)f_2$. This is positive if $\epsilon < (\gamma - \beta_1)(f_2/f_1)$. We can now conclude that the model allows for an upward-sloping demand curve if we choose the parameters $\beta_1, \epsilon, \gamma, f_1$ and $f_2$ to be positive and so that:

$$\frac{\gamma - \beta_1}{\beta_1} < \epsilon < (\gamma - \beta_1)\frac{f_1}{f_2}. \tag{41}$$

This makes clear that the model developed here indeed allows for upward-sloping demand curves for a range of possible parameter values.

Note that the model derived in this section is even closer to the standard model of consumer demand in the two good case than that of the previous section in that we consider the demand for a homogeneous good. The only difference with the standard model is that the total demand for this commodity is derived from two 'sources': the demand for kilometres for the two cars. Although at least one of the demand functions for kilometres of the individual cars is downward sloping, the aggregate demand function may be upward-sloping. This positive net effect can, of course, not be interpreted as caused by substitution in the conventional sense of that word, since we would then expect the total number of kilometers to decrease or, at most, to remain unchanged, as has been discussed above. However, economic theory allows for the possibility that in this case the effect of the price increase on total demand is positive.

# 4 Quality Choice and Long Run Demand for Using a Durable

## *4.1 A Model for Quality Choice of a Durable Good*

The third situation to be discussed in this paper concerns the long run demand function for the services provided by a durable good. Probably the best example to think of is that of a private car.

To study quality choice, we must take into account that quality will affect both fixed and variable costs. Quality will be treated as a continuous variable and fixed as well as variable car costs are assumed to be increasing functions of car quality, to be denoted as $F(k)$ and $p(k)$, respectively.

To investigate quality choice, we take the conditional indirect utility function as our starting point:

$$v = v(p(k), y - F(k), k). \tag{42}$$

This utility function gives the maximum level of utility the consumer is able to reach while taking the quality of the car as fixed. Car quality affects indirect utility in three ways: directly (as an argument of the indirect utility function) and indirectly via the fixed and variable costs. The consumer chooses the quality of this car, as well as the number of kilometres to be driven with it. The latter follows in the usual way via Roy's identity. The first order condition for maximization with respect to $k$ is:

$$\frac{\partial v}{\partial(y-F)}\frac{\partial F}{\partial k}+\frac{\partial v}{\partial k}+\frac{\partial v}{\partial p}\frac{\partial p}{\partial k}=0. \tag{43}$$

After rearranging terms and application of Roy's identity, we can rewrite this as:

$$wtp(k)=\frac{\partial F}{\partial k}+q\frac{\partial p}{\partial k}, \tag{44}$$

where $wtp(k)$ denotes the marginal willingness to pay for car quality:

$$wtp(k)=\frac{\partial v/\partial k}{\partial v/\partial(y-F)}. \tag{45}$$

The first order condition can thus be interpreted as saying that the marginal willingness to pay for car quality equals the additional fixed and variable costs associated with a marginal increase in quality. Alternatively, we can rewrite (43) as:

$$\frac{\partial F}{\partial k}=q\left(mrs-\frac{\partial p}{\partial k}\right) \tag{46}$$

where *mrs* denotes the marginal rate of substitution between the variable cost and the quality of the car:

$$mrs=\frac{\partial v/\partial k}{\partial v/\partial p}. \tag{47}$$

This marginal rate of substitution gives the increase in variable cost that the household is willing to accept in exchange for an increase in car quality. According to (46), this quantity must exceed the effect of the higher quality on variable cost. The expression in parentheses in the right hand side of (46) can thus be interpreted as the net marginal utility of higher quality per kilometre driven. Multiplication of this change in surplus by the number of car kilometres must result in a value equal to the effect of the increase in quality on the fixed cost. If follows from (46) that the long run demand equation must satisfy:

$$q^{lr}=\frac{\frac{\partial F}{\partial k}}{mrs-\frac{\partial p}{\partial k}}. \tag{48}$$

We expect the long run demand equation to be decreasing in the fuel price $\pi$. However, it is not obvious from (48) that this will indeed be the case. A particular reason for doubt in this respect is that the marginal variable cost of quality $\partial p/\partial k$ is probably increasing in the fuel price, whereas the other variables on the right-hand side of (48) have no obvious relationship with the fuel price. This suggests that the long run demand equation may well be upward-sloping.

## 4.2 An Example: Car Quality Choice

We use a model that is similar to that of Sect. 3.2, but concerns a household that owns just one car. Indirect utility is then given by[5]:

$$v = g(y - F) + \frac{1}{2}\frac{(\alpha - p)^2}{\beta}. \tag{49}$$

To analyze quality choice, we treat car quality as a continuous nonnegative variable and assume that the parameter $\alpha$ is linearly increasing in car quality:

$$\alpha = \alpha_0 + \alpha_k k \tag{50}$$

For fixed cost $F$ and variable cost $p$ we make similar assumptions:

$$F = F_0 + F_k k \tag{51}$$

$$p = (v_0 + v_k k)\pi. \tag{52}$$

The expression between brackets on the right-hand-side of (52) gives the amount of gasoline used per kilometre driven (that is, the fuel intensity to which we referred as $f$ in the previous section). It is multiplied by the price $\pi$ of gasoline to arrive at the variable cost.

In what follows we assume that $g$ in (49) is quadratic:

$$g(y) = y - \delta y^2, \qquad \delta \geq 0. \tag{53}$$

After making the substitutions suggested by (50), (51), (52) and (53), (49) can be rewritten as:

$$v = y - F_0 - F_k k - \delta(y - F_0 - F_k k)^2 + \frac{(\alpha_0 + \alpha_1 k - (v_0 + v_k k)\pi)^2}{2\beta}, \tag{54}$$

$$= y - F_0 - \delta(y - F_0)^2 - dF_k k - \delta F_k^2 k^2 + \frac{(a + bk)^2}{2\beta}. \tag{55}$$

In (55) the following notation has been introduced:$d = (1 - 2\delta(y - F_0))$, $a = \alpha_0 - v_0\pi$, and $b = \alpha_k - v_k\pi$. The variable $d$ is the marginal utility valuated at $y - F_0$, and must clearly be positive; $a$ is the value of $(\alpha - p)$ when car quality is equal to zero, and we assume this to be positive as well; $b$ gives the effect of an increase in car quality on the utility generated by the first kilometre driven, and must be positive.

---

[5]This indirect utility function is a special case of the one studied in the previous section.

Equation (55) is quadratic in $k$. Optimal quality choice follows from the first order condition (43):

$$k = \frac{ab - \beta d}{2\delta F_k^2 \beta - b^2}. \tag{56}$$

The second order condition for this maximization problem requires that the denominator of the right-hand-side of (56) is positive. We assume that the numerator is also positive. Given these assumptions, it can be verified that the optimal value of $k$ is increasing in income, $y$, and in the intensity of preference for car 1, $\alpha_k$, as one would expect. It is decreasing in the two parameters, $F_0$ and $F_k$ of the fixed cost function, in the price of gasoline $\pi$, in the two coefficients, $v_0$ and $v_k$ of the variable cost function and in and $\delta$ and $\beta_1$.

The short run demand for kilometres can be derived from (49) by Roy's identity. If we rewrite it in the notation just introduced it is:

$$q^{sr} = \frac{a + bk}{\beta(d - 2\delta F_k k)}. \tag{57}$$

Its partial derivative with respect to the price of gasoline is:

$$\frac{\partial q^{sr}}{\partial \pi} = -\frac{v_o + v_k k}{\beta(d - 2\delta F_k k)} \tag{58}$$

which is negative, as expected. Substitution of the equation for optimal car quality (56) into the short run demand function, or application of (48), gives the long run demand function:

$$q^{lr} = \frac{F_k}{b} = \frac{F_k}{\alpha_k - v_k \pi} \tag{59}$$

which is a surprisingly simple formulation. However, it has the immediate implication that long run demand is increasing in the gasoline price $\pi$.

## 5 Conclusion

This paper has considered the effects of substitution within a group of commodities on the demand for the total number of units of that group. It has identified three specific cases in which the latter demand curve may be upward-sloping and provided concrete examples in which this occurs for specific parameter values. Since all three examples work in the absence of negative income effects, the mechanism behind these upward-sloping demand curves is clearly different from that of the traditional Giffen goods. Taken together, the examples suggest that our intuitive notion of the concept of substitution differs in some respect from what happens when a utility maximizing consumers is confronted with simultaneous increases in the prices of close substitutes. Since there appear to be many examples of such close substitutes, concerning durables as well as non-durables, the finding of the present paper may well be relevant for empirical work.

**Acknowledgements** The author thanks two anonymous referees and Pierre von Mouche for useful comments on an earlier version and Thomas de Graaff for typesetting the latex version. The usual disclaimer applies.

## References

1. Blundell, R. and Robin, J. M.: 2000, Latent separability: Grouping goods without weak separability, *Econometrica* **68**, 53–84.
2. Deaton, A. and Muellbauer, J.: 1980, *Economics and Consumer Behavior*, Cambridge University Press, Cambridge.
3. Dubin, J. A. and McFadden, D.: 1984, An Econometric Analysis of Residential Electric Appliance Holdings and Consumption, *Econometrica* **52**, 345–362.
4. Hausman, J. A.: 1979, Individual Discount Rates and the Purchase and Utilization of Energy-Using Durables, *Bell Journal of Economics* **10**, 33–54.
5. Jensen, R. T. and Miller, N. H.: 2008, Giffen Behavior and Subsistence Consumption, *American Economic Review* **98**, 1553–1577.
6. LaFrance, J. T.: 2008, The structure of US food demand, *Journal of Econometrics* **147**, 336–349.
7. Mas Colell, A., Whinston, M. D. and Green, J. R.: 1995, *Microeconomic theory*, Oxford University Press, Oxford.
8. Rosen, S.: 1974, Hedonic Prices and implicit Markets: Product Differentiation in Pure Competition, *Journal of Political Economy* **82**, 34–55.

# Lotteries and the Law of Demand

**Rodney J. Garratt**

**Abstract** In economies with nonconvexities consumers can increase their expected utility by consuming lotteries. Lotteries are probability distributions over bundles in the consumption set. Standard revealed preference logic can be applied to choices in lottery space, however the implications are not readily interpretable. In this paper, we formulate the law of demand for lottery economies in terms of commodity price changes and changes in demand for commodities. The finding is that the standard expression of the compensated law of demand necessarily holds in expectation only.

## 1 Introduction

In economies with nonconvexities consumers sometimes prefer to have their final consumption determined by lottery. By randomizing over consumption bundles, even a risk averse consumer can achieve higher expected utility than if she chooses any single consumption bundle without randomizing. This point was first made in the context of indivisible goods in [11]. Discussions of how lotteries improve consumer welfare in situations where consumers make indivisible expenditures on education and improvements to property are found in [7] and [8], respectively.

Lottery demands are probability distributions over the consumption bundles in the consumption set. When lottery demands satisfy the weak axiom of revealed preference the implication is that for compensated price changes

$$(\Pi' - \Pi) \cdot (L' - L) \leq 0, \tag{1}$$

where $\Pi$ and $\Pi'$ are two price vectors that have dimension equal to the number of available commodity bundles, and $L$ and $L'$ are the corresponding lottery demands.

R.J. Garratt (✉)
Department of Economics, University of California, Santa Barbara, CA 93106, USA
e-mail: garratt@econ.ucsb.edu

W. Heijman and P. Mouche (eds.), *New Insights into the Theory of Giffen Goods*, Lecture Notes in Economics and Mathematical Systems 655,
DOI 10.1007/978-3-642-21777-7_12, 

The law of demand in this setting, which can be expressed as a special case of (1) where $\Pi$ and $\Pi'$ differ by a single component, says that the probability demanded on a commodity bundle must vary inversely with a compensated change in the price of the bundle.

This statement of the law of demand is somewhat removed from what people (economists and actual consumers) normally focus on, namely, changes in prices of commodities and changes of demands for commodities. In this paper, the law of demand for lottery economies is expressed in terms of commodity price changes and changes in demand for commodities. This is done by mapping lotteries into state-contingent consumptions, so that the resulting state-contingent demand vectors, $x(s)$, have dimension equal to the number of commodities, not the number of commodity bundles.

Suppose $p$ and $p'$ are two vectors of prices of individual commodities. The main result is that satisfaction of the weak axiom in lottery economies implies that

$$\int_S (p' - p) \cdot (x'(s) - x(s))ds \leq \mathbf{0}. \tag{2}$$

Thus, in suitably defined lottery economies, the standard expression of the law of demand, $(p' - p) \cdot (x' - x) \leq 0$, necessarily holds in expectation only.

In Sect. 2, we establish the main result for the case where the commodity set has a finite number of bundles. The literature on lottery equilibrium often assumes that the consumption set has only a finite number of points because this simplifies notation, proofs, and computations (see [3], [6], [12], and [13]). However, the main result extends quite easily to more general commodity spaces with an uncountable number of commodities (see Remark 2). Section 3 provides an example where the consumer selects consumption bundles comprised of 0 or 1 units of a single indivisible good and any quantity of a divisible one. The case of $\{0, 1\}$ goods has received much attention in the literature on labor-market lotteries ([9], [14], [15]). The example describes a scenario in which the probability demanded on the indivisible good increases when its price rises. Nevertheless, it is easily verified that the lottery law of demand, and its alternative expression provided in (5) below, holds when income is compensated for the price change. A brief discussion of the main result is provided in Sect. 4.

## 2 Model and Results

A consumer with endowed wealth $w$ is deciding on her consumption of $n$ goods. Her consumption set $C$ is nonconvex, and hence she considers lotteries. In this section, we assume that she only has access to a finite set of possible consumption bundles, so her consumption set $C$ is finite, $|C| = m$. Thus, we consider lottery demands $L = (q_1, \ldots, q_m)$, where $q_i$ denotes the probability demanded on commodity bundle $i$, $\sum_{i=1}^{m} q_i = 1$ and $q_i \geq 0$.

Suppose the cost of lottery $L$ is given by $\Pi \cdot L$ where $\Pi \in \Re^m_+$. Suppose that for given prices $\Pi$ and wealth $w$ the consumer has lottery demand $L(\Pi, w)$ that is homogeneous of degree zero and satisfies Walras' Law.

**Definition 1.** (WARP): The lottery demand function $L(\Pi, w)$ satisfies the weak axiom of revealed preference if for any two price-wealth situations $(\Pi, w)$ and $(\Pi', w')$ : If $\Pi \cdot L(\Pi', w') \leq w$ and $L(\Pi', w') \neq L(\Pi, w)$ then $\Pi' \cdot L(\Pi, w) > w'$.

WARP implies that for compensated price changes

$$(\Pi' - \Pi) \cdot [L(\Pi', w') - L(\Pi, w)] \leq 0 \tag{3}$$

with strict inequality whenever $L(\Pi', w') \neq L(\Pi, w)$ (See, for example, p. 31, part (i) of [10]). Note that for reasons outlined in [3] no arbitrage implies there exists $p \in \Re^n_+$ such that $\Pi_i = p \cdot c_i$. Thus, (3) becomes (omitting obvious transposes)

$$\begin{pmatrix} p' \cdot c_1 - p \cdot c_1 \\ \vdots \\ p' \cdot c_m - p \cdot c_m \end{pmatrix} \cdot [L(\Pi', w') - L(\Pi, w)] \leq 0$$

or

$$\begin{pmatrix} (p' - p) \cdot c_1 \\ \vdots \\ (p' - p) \cdot c_m \end{pmatrix} \cdot [L(\Pi', w') - L(\Pi, w)] \leq 0. \tag{4}$$

Fix a continuous probability space $(S, \sum, \mu)$ where $S = [0, 1]$, $\sum$ is the Borel sets on $S$ and $\mu$ is the Lebesgue measure. By Lemma 4 of [5] there exists $x : S \to \Re^n_+$ such that

$$L(\Pi, w) = \mu \circ x^{-1}.$$

This establishes that, given a lottery $L$, one can find a state contingent consumption $x$ defined over $S$ with the property that for each bundle $c_i$, $i = 1, \ldots, m$, the measure of the set of states $\{s : x(s) = c_i\}$ is $q_i$.

Given this result, (4) becomes

$$\begin{pmatrix} (p' - p) \cdot c_1 \\ \vdots \\ (p' - p) \cdot c_m \end{pmatrix} \cdot [\mu \circ x'^{-1} - \mu \circ x^{-1}] \leq 0$$

or

$$\begin{pmatrix} (p' - p) \cdot c_1 \\ \vdots \\ (p' - p) \cdot c_m \end{pmatrix} \cdot \begin{pmatrix} \int_{S_1'} ds - \int_{S_1} ds \\ \vdots \\ \int_{S_m'} ds - \int_{S_m} ds \end{pmatrix} \leq 0$$

where $\int_{S_i} ds = \int_{S:x(s)=c_i} ds$. Rewrite the left-hand-side to get

$$(p'-p)\cdot\left[c_1\left(\int_{S_1'} ds - \int_{S_1} ds\right) + \cdots + c_m\left(\int_{S_m'} ds - \int_{S_m} ds\right)\right]$$

$$= (p'-p)\cdot\left[\left(\int_{S_1'} c_1 ds + \cdots + \int_{S_m'} c_m ds\right) - \left(\int_{S_1} c_1 ds + \cdots + \int_{S_m} c_m ds\right)\right]$$

$$= (p'-p)\cdot\left[\left(\int_{S_1'} x_1'(s) ds + \cdots + \int_{S_m'} x'(s) ds\right)\right.$$

$$\left. - \left(\int_{S_1} x(s) ds + \cdots + \int_{S_m} x(s) ds\right)\right] = (p'-p)\cdot\left[\int_0^1 x'(s) ds - \int_0^1 x(s) ds\right].$$

Thus we have

$$\int_0^1 (p'-p)\cdot(x'(s) - x(s)) ds \leq 0.$$

For a change in the price of a single commodity $i$ this becomes

$$\int_0^1 (p_i' - p_i)(x_i'(s) - x_i(s)) ds \leq 0, \tag{5}$$

which is the desired result.

*Remark 1.* Suppose $L$ is such that $q_i$ is a rational number for $i = 1, \ldots, m$. Then we can duplicate the above process using a finite set of $k$ equally-probable extrinsic states to get

$$\sum_{j=1}^{k} (p'-p)\cdot(x'(s_j) - x(s_j)) \leq 0.$$

*Remark 2.* Redefine the consumption set $C$ to be any (possibly nonconvex) subset of $\Re^n_+$. Let $\Delta(C)$ denote the set of probability measures over the set C. An individual lottery demand is an element of $\Delta(C)$, $\delta : B(C) \to \Re_+$, where $B(C)$ denotes the Borel subsets of a given set C. We have assumed that lottery prices are such that a unit of probability on commodity bundle $c \in C$ costs $p \cdot c$. Thus we can define an optimally chosen lottery, given commodity prices $p$ and income $w$, by $\delta(p, w)$.[1]

[1] For a discussion of the consumer's lottery choice problem in this environment see [5].

A wide class of examples will involve solutions to the lottery choice problem that have countable support. For instance, this will be true if $C = Z_+^{\ell_1} \times \Re_+^{\ell_2}$ where $\ell_1$ is the number of indivisible goods and $\ell_2$ is the number of divisible goods, and her utility function is strictly concave on $\Re_{++}^{\ell_2}$. Risk averse consumers will not engage in unnecessary randomization, and hence the support of the chosen distribution will typically be quite small.

Suppose that the consumer always demands strictly positive probability on a finite number of commodity bundles. Then, given two price-income combinations $p, w$ and $p', w'$, let $\{c_1, \ldots, c_m\}$ denote the joint list of commodity bundles that receive strictly positive probability by the consumer in either of her selected lotteries. Let $L = \{q_1, \ldots, q_m\}$ and $L' = \{q'_1, \ldots, q'_m\}$ denote $m$-dimensional vectors where $q_i = \delta(c_i; p, w)$ and $q'_i = \delta(c_i; p', w')$, for $i = 1, \ldots, m$. Note that for each vector $L$ and $L'$ the components are non-negative and sum to 1. The statement of WARP and the implications for compensated price changes can be evaluated using these finite-vector expressions of lottery demands, as shown in Sect. 3.1.

## 3 Example

Consider a consumer with endowed wealth $w$ choosing consumption of an indivisible good $x_1 \in \{0, 1\}$ and perfectly divisible (composite) good $x_2$. Suppose utility for the consumer is represented by a von Neumann–Morgenstern utility function $u = U(x)$. $U$ is twice differentiable with respect to $x_2$, $U(1, x_2) > U(0, x_2)$ for all $x_2$, and $U$ is strictly increasing and strictly concave in $x_2$. Suppose the price of the indivisible good is $p_1$ and the price of the divisible good is 1. Any money that is not spent on the indivisible good is spent on the divisible good. Thus, without lotteries, and provided $p_1 < w$, the consumer's problem is simply a choice between two alternatives, $(0, w)$ and $(1, w - p_1)$. The former solution occurs whenever $U(0, w) > U(1, w - p_1)$ while the latter occurs whenever $U(1, w - p_1) > U(0, w)$. If $U(1, w - p_1) = U(0, w)$ then both $(0, w)$ and $(1, w - p_1)$ are solutions.

Using $z$ as a running variable for wealth, indirect utility is denoted by

$$F(z; p) = \max[U(0, z), U(1, z - p_1)],$$

and is shown as the upper envelope of the two utility curves in Fig. 1. The key to understanding the figure is to recognize that consuming the indivisible good requires an expenditure of $p_1$. Thus the upper curve in Fig. 1, denoted $U(1, z - p_1)$, is just the curve $U(1, z)$ shifted to the right by the cost of the indivisible good. Given our assumptions on utility it is easy to establish the existence of a reservation price for the indivisible good, below which the indivisible good is always consumed and above which it is not. The existence of a reservation price implies that according to the standard analysis the indivisible good cannot be a Giffen good. Because of the indivisibility, however, consumers may increase their utility exante by randomizing

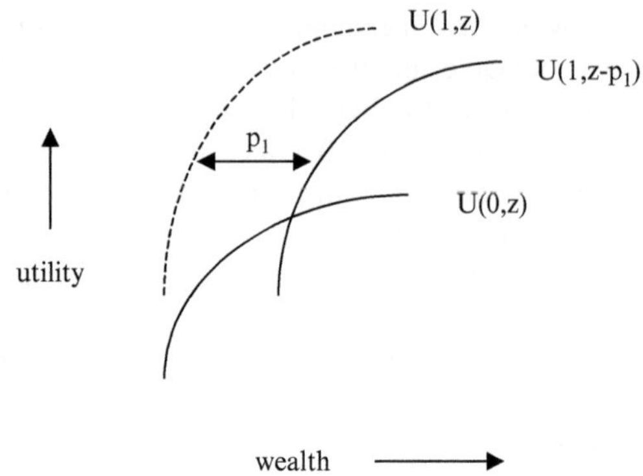


**Fig. 1** Indirect utility

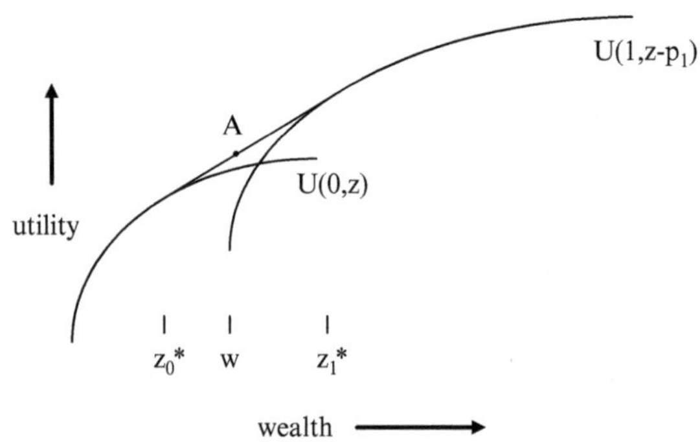


**Fig. 2** Optimum gamble

over final consumption bundles. This is because indivisibility leads to risk loving by consumers.

The indirect utility shown in Fig. 1 is of the form posed by Friedman and Savage [2]. Thus the consumer can gain from a gamble. This is shown in Fig. 2. To obtain Fig. 2, take Fig. 1 and form the least concave hull by drawing a straight line that is just tangent to the two utility curves. Point $z_0^*$ is the left tangency on the curve $U(0,z)$. Point $z_1^*$ is the right tangency on the curve $U(1,z-p_1)$. Suppose endowed wealth $w$ lies somewhere between $z_0^*$ and $z_1^*$. Point $A$ is on the linear segment used to form the convex hull of the indirect utility frontier, directly above $w$. It is apparent from Fig. 2 that the consumer obtains her highest possible expected utility at point $A$ by trading all of her wealth for a gamble in money that pays either $z_1^*$ or $z_0^*$. If the gamble pays $z_1^*$ then she purchases the indivisible good. If it pays $z_0^*$, she does not. The payoff $z_1^*$ is defined to be the wealth level that corresponds to the decision to purchase the indivisible good. Thus, the probability of receiving the payoff $z_1^*$, to be denoted by $\lambda^*$, is the expected demand for the indivisible good.

Actuarial fairness implies that

$$\lambda^* = \frac{w - z_0^*}{z_1^* - z_0^*}. \tag{6}$$

Assuming the consumer is an expected utility maximizer, her optimization problem is written formally as

$$\max_{\lambda, z_0, z_1} (1-\lambda)U(0, z_0) + \lambda U(1, z_1 - p_1)$$

subject to

$$\lambda = \frac{w - z_0}{z_1 - z_0}$$
$$0 \le \lambda \le 1$$
$$z_0 \ge 0,\ z_1 \ge 0.$$

The first order conditions for an interior solution are given by (6),

$$U_2(0, z_0^*) = U_2(1, z_1^* - p_1), \tag{7}$$

and

$$U(1, z_1^* - p_1) = U(0, z_0^*) + U_2(0, z_0^*)[z_1^* - z_0^*]. \tag{8}$$

Despite the non-concavity of the objective function and the nonlinearity of the constraint (6), the values of $z_0^*$, $z_1^*$ and $\lambda^*$ that solve the first order conditions represent a global maximum (see [7]).

Interior solutions occur when prices and endowed wealth are such that $w$ lies in a nonconcave region of $F(z; p)$. If prices and endowed wealth are such that $F(z; p)$ is concave in $z$ at $w$ and $F(z; p) = U(0, w)$ then $\lambda^* = 0$ is chosen and $x_2 = w$ is the consumption of the divisible good. If prices and endowed wealth are such that $F(z; p)$ is concave at $w$ and $F(w; p) = U(1, w - p_1)$ then $\lambda^* = 1$ is chosen and $x_2^* = w - p_1$ is the consumption of the divisible good.

It is now possible to analyze how the optimum $\lambda$ responds to changes in $p_1$. Let $R(0, z_0) = -U_{22}(0, z_0)/U_2(0, z_0)$ and $R(1, z_1) = -U_{22}(1, z_1 - p_1)/U_2(1, z_1 - p_1)$ denote the Arrow–Pratt measures of absolute risk aversion at points on each of the curves, $U(0, z_0)$ and $U(1, z_1 - p_1)$ respectively. Then, from the first order equations,

$$\partial\lambda^*/\partial p_1 = \frac{-[(1-\lambda^*)/R(0, z_0^*) + \lambda^*/R(1, z_1^*) + \lambda^*(z_1^* - z_0^*)]}{(z_1^* - z_0^*)^2}. \tag{9}$$

The sign of (9) can be positive. Thus there can be a positive price effect on the expected demand for the indivisible good.

For (9) to be positive it is necessary that $z_1^* - z_0^* < 0$. Clearly, this is not the case for the consumer with indirect utility drawn in Fig. 2. Rather, endowed wealth must fall in a nonconcave region of $F(z; p)$ where $U(0, z)$ cuts $U(1, z - p_1)$ from below. This is shown in the Fig. 3. To achieve the situation shown in Fig. 3, the curve $U(1, z - p_1)$ must cross the curve $U(0, z)$ more than once. In instances where $U(1, z - p_1)$ crosses the curve $U(0, z)$ from above the potential exists for a Giffen effect. The consumer with endowed wealth $w$ in Fig. 3 obtains maximum expected

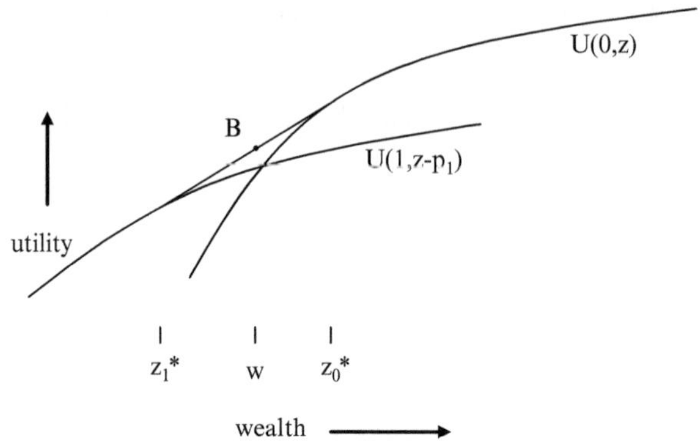


**Fig. 3** Upward sloping demand

utility at point $B$. The payoffs in the optimum gamble for this consumer satisfy $z_1^* - z_0^* < 0$.

Cook and Graham [1] defines the indivisible good to be inferior if its reservation price is a decreasing function of income. This will be true at wealth levels near a point where the curve $U(0, z)$ cuts the curve $U(1, z - p_1)$ from below. Thus, at endowed wealth $w$, the indivisible good is an inferior good. Not surprisingly, inferiority of the indivisible good is only a necessary condition for (9) to have a positive sign.

Next, an example is presented in which the probability demanded by the consumer for the indivisible good varies directly with own price. Suppose the consumer's utility function is given by

$$U(0, x_2) = \begin{cases} \left[ \left( \frac{7.425}{8.9} - .75 \right)^{.1} \Big/ \left( \frac{7.425}{8.9} \right)^{.99} \right] x_2^{.99} \text{ if } 0 \leq x_2 \leq \frac{7.425}{8.9} \\ (x_2 - .75)^{.1} \text{ if } x_2 > \frac{7.425}{8.9} \end{cases}$$

and $U(1, x_2) = .4 + .5x_2^{.9}$. Using two pieces to construct the function $U(0, x_2)$ is a convenient way to make the term $z_1^* - z_0^*$ in (9) as large a negative number as possible, while maintaining large positive values of absolute risk aversion at each of the solution points.[2] The function $U(0, \cdot)$ is, by construction, continuous and differentiable.

Suppose endowed wealth is $w = 0.5$ and the price of the divisible good is 1. We are interested in how $\lambda^*$ varies with $p_1$. For $p_1 \leq 0.0744$, the indirect utility function $F(z; p)$ is concave at $w = 0.5$ and $F(0.5; p) = U(1, 0.5 - p_1)$. Thus, for $p_1 \leq 0.0744$, $\lambda^* = 1$. For $p_1 \geq 0.42451$, the indirect utility function $F(z; p)$ is again concave at $w = 0.5$ only now $F(0.5; p) = U(0, 0.5)$. Thus, for $p_1 \geq 0.42451$,

[2] An example constructed from a single functional form is provided in Garratt [4].

$\lambda^* = 0$. For values of $p_1$ in between 0.0744 and 0.42451, $F(z; p)$ is convex at $w = 0.5$, and $\lambda^*$ is obtained by solving equations (7) and (8) with the given functional forms to get $z_0^*$ and $z_1^*$, and then substituting these values into (6).

It is not possible to obtain closed form solutions for $z_0^*$ and $z_1^*$. Therefore, the solutions are obtained numerically using Newton's method. Over a range of prices corresponding to an interior solution, $\lambda^*$ varies directly with its own price. For example at $p_1 = 0.2$, $\lambda^* \approx 0.607307$ and at $p_1' = 0.4$, $\lambda^* \approx 0.7\dot{7}$. In between those prices $\lambda^*$ increases monotonically with $p_1$.

## 3.1 Verification of the Lottery Law of Demand

Consider the commodity price vectors $p = (0.2, 1)$ and $p' = (0.4, 1)$ from the above example. At $p_1 = 0.2$, $z_0^*(p) = 0.885084$ and $z_1^*(p) = 0.250100$. So $z_1^*(p) - p_1 = 0.050100$. At $p_1' = 0.4$, $z_0^*(p') = 0.839582$ and $z_1^*(p') = 0.401265$. So $z_1^*(p') - p_1' = 0.001265$. Assuming an interior solution, the optimal lottery $\delta$ corresponding to commodity prices $p$ and income $w$ is

$$\delta(c; p, w) = \begin{cases} \left(1 - \dfrac{w - z_0^*(p)}{z_1^*(p) - z_0^*(p)}\right) & \text{if } c = (0, z_0^*(p)) \\ \dfrac{w - z_0^*(p)}{z_1^*(p) - z_0^*(p)} & \text{if } c = (1, z_1^*(p) - p_1) \\ 0 & \text{otherwise} \end{cases}$$

As indicated in Remark 2, we can express these lotteries in terms of finite vectors, $L$ and $L'$, and demonstrate that (3) holds. Since there are only two distinct commodity bundles that receive positive probability for each set of commodity prices and income, we can evaluate the implications of WARP using vectors with four components. Let $c_1 = (0, z_0^*(p)) = (0, 0.885084)$, $c_2 = (1, z_1^*(p) - p_1) = (1, 0.050100)$, $c_3 = (0, z_0^*(p')) = (0, 0.839582)$ and $c_4 = ((1, z_1^*(p') - p_1') = (1, 0.001265)$. Define

$$\Pi = c \cdot p = \begin{cases} 0.885084 \text{ if } c = c_1 \\ 0.250100 \text{ if } c = c_2 \\ 0.839582 \text{ if } c = c_3 \\ 0.201265 \text{ if } c = c_4 \end{cases}$$

and

$$\Pi' = c \cdot p' = \begin{cases} 0.885084 \text{ if } c = c_1 \\ 0.450100 \text{ if } c = c_2 \\ 0.839582 \text{ if } c = c_3 \\ 0.401265 \text{ if } c = c_4 \end{cases}$$

So $\Pi' - \Pi = (0, 0.2, 0, 0.2)$.

Let

$$L = (1 - \frac{w - z_0^*(p)}{z_1^*(p) - z_0^*(p)}, \frac{w - z_0^*(p)}{z_1^*(p) - z_0^*(p)}, 0, 0) = (0.392693, 0.607307, 0, 0)$$

denote the finite vector corresponding to the lottery demand $\delta(p, w)$. The change in income required to compensate the consumer for the price change from $\Pi$ to $\Pi'$ is $\Delta w = (\Pi' - \Pi) \cdot L = 0.121461$. Setting $w' = 0.5 + \Delta w = 0.621461$ we get that $\frac{w' - z_0^*(p')}{z_1^*(p') - z_0^*(p')} = \frac{0.621461 - 0.839582}{0.401265 - 0.839582} = 0.497633$.

Let

$$L' = (0, 0, 1 - \frac{w' - z_0^*(p')}{z_1^*(p') - z_0^*(p')}, \frac{w' - z_0^*(p')}{z_1^*(p') - z_0^*(p')}) = (0, 0, 0.502367, 0.497633)$$

denote the finite vector corresponding to the lottery demand $\delta(p', w')$. So $L' - L = (-0.392693, -0.607307, 0.502367, 0.497633)$. Thus $(\Pi' - \Pi) \cdot (L' - L) = -0.021935$ which is less than zero, as required.

To demonstrate the main result let

$$x(s) = \begin{cases} c_1 \text{ if } s \in [0, 0.392693) \\ c_2 \text{ if } s \in [0.392693, 1] \end{cases}$$

and

$$x'(s) = \begin{cases} c_3 \text{ if } s \in [0, 0.502367) \\ c_4 \text{ if } s \in [0.502367, 1] \end{cases}$$

Then we have $p_1' - p_1 = 0.2$ and

$$x_1'(s) - x_1(s) = \begin{cases} 0 & \text{if } s \in [0, 0.392693) \\ -1 & \text{if } s \in [0.392693, 0.502367) \\ 0 & \text{if } s \in [0.502367, 1] \end{cases}$$

So (5) becomes

$$\int_0^1 (p_1' - p_1)(x_1'(s) - x_1(s))ds = \int_{0.392693}^{0.502367} 0.2(-1)ds = -0.021935 \leq 0.$$

## 4 Discussion

The practical implication of the analysis is conveyed by the following scenario. Each year Alice undertakes a gamble to determine whether or not to take a trip to Hawaii. The odds of the gamble depend on how much money she "bets" and the price of the trip. When trips are cheap she can get better odds for the same money than when trips are expensive. Nevertheless, the example shows that she may increase her probability of going to Hawaii when the price increases. This means that empirically, we could observe her going to Hawaii more often in years when the price is high than when it is low. Of course, many years of observations would be required to establish upward sloping demand. Even if Alice chooses a lower probability of going to Hawaii when the price rises, it is possible that she will lose the more-favorable gamble the first year and win the less-favorable gamble the second year. However, the law of demand does not rule out the possibility of upward sloping demand curves in actual quantities or in consumption of probabilities. The law requires that when price rises, Alice's income compensated demand is necessarily lower. This is established by the main result and demonstrated for the case of upward sloping (uncompensated) demand by the example.

## References


1. Cook, P.J. and D.A. Graham, "The Demand for Insurance and Protection: The Case of Irreplaceable Commodities," *Quarterly Journal of Economics*, 1977, 91, 143–156.
2. Friedman, M. and L. Savage, "Utility Analysis of Choices Involving Risk," *Journal of Political Economy*, 1948, 56(4), 279–304.
3. Garratt, R., "Decentralizing Lottery Allocations in Markets With Indivisible Commodities, *Economic Theory*, 1995, 5, 295–313.
4. Garratt, R., "A Tale of Two Cities and a Giffen Good," *Canadian Journal of Economics*, February 2005, 38(1), 49–56.
5. Garratt, R., T. Keister, C.-Z. Qin and K. Shell, "Equilibrium Prices when the Sunspot Variable is Continuous, *Journal of Economic Theory*, November 2002, 107, 11–38.
6. Garratt, R., T. Keister and K. Shell, "Comparing Sunspot Equilibrium and Lottery Equilibrium Allocations: The Finite Case, *International Economic Review*, May 2004, 45(2),351–386.
7. Garratt, R. and J.M. Marshall, "Public Finance of Private Goods: The Case of College Education, *Journal of Political Economy*, 1994, 102(3), 566–582.
8. Garratt, R. and J.M. Marshall, "Insurable Interest, Options to Convert and Demand for Upper Limits in Optimum Property Insurance," *Journal of Risk and Insurance*, 1996, 63, 185–206.
9. Hansen, G., "Indivisible Labor and the Business Cycle, *Journal of Monetary Economics*, 1985, 16, 309–328.
10. Mas-Colell, A., M.D. Whinston and J.R. Green, 1995, Microeconomic Theory, Oxford University Press, New York.
11. Ng, Y.-K., "Why do People Buy Lottery Tickets? Choices Involving Risk and the Indivisibility of Expenditure," *Journal of Political Economy*, 1965, 73, 530–535.
12. Prescott, E. and R. Townsend, "General Competitive Analysis in an Economy with Private Information," *Internation Economic Review*, 1984, 25, 1–20.
13. Prescott, E. and R. Townsend, "Firms as Clubs in Walrasian Markets with Private Information, Working Paper 00-8, Federal Reserve Bank of Richmond, September 2000.
14. Rogerson, R., "Indivisible Labor, Lotteries and Equilibrium," *Journal of Monetary Economics*, 1988, 21, 3–16.
15. Shell, K. and R. Wright, "Indivisibilities, Lotteries, and Sunspot Equilibria, *Economic Theory*, 1993, 3, 1–17.

MIX
Papier aus verantwortungsvollen Quellen
Paper from responsible sources
FSC® C105338

If you have any concerns about our products,
you can contact us on
**ProductSafety@springernature.com**

In case Publisher is established outside the EU,
the EU authorized representative is:
**Springer Nature Customer Service Center GmbH**
**Europaplatz 3, 69115 Heidelberg, Germany**

Printed by Libri Plureos GmbH
in Hamburg, Germany